Communications
Network Analysis

Arthur D. Little Books

A series of books on management and other scientific and technical subjects by senior professional staff members of Arthur D. Little, Inc., the international consulting and research organization. The series also includes selected nonproprietary case studies.

Acquisition and Corporate Development
   James W. Bradley and Donald H. Korn

Bankruptcy Risk in Financial Depository Intermediaries: Assessing Regulatory Effects
   Michael F. Koehn

Board Compass: What It Means to Be a Director in a Changing World
   Robert Kirk Mueller

Career Conflict: Management's Inelegant Dysfunction
   Robert Kirk Mueller

Communications Network Analysis
   Howard Cravis

The Corporate Development Process
   Anthony J. Marolda

Corporate Responsibilities and Opportunities to 1990
   Ellen T. Curtiss and Philip A. Untersee

The Dynamics of Industrial Location: Microeconometric Modeling for Policy Analysis
   Kirkor Bozdogan and David Wheeler

Energy Models for Forecasting and Policy Analysis
   Glenn R. De Souza

The Incompleat Board: The Unfolding of Corporate Governance
   Robert Kirk Mueller

System Methods for Socioeconomic and Environmental Impact Analysis
   Glenn R. De Souza

Communications Network Analysis

Howard Cravis
Arthur D. Little, Inc.

An Arthur D. Little Book

LexingtonBooks
D.C. Heath and Company
Lexington, Massachusetts
Toronto

To

Susi, Bill, and Steven

Library of Congress Cataloging in Publication Data

Cravis, Howard.
  Communications network analysis.

  "An Arthur D. Little book."
  Includes bibliographical references.
  1. Telecommunications systems. 2. Electric network analysis. I. Title.
TK5103.C7   621.38   75-39314
ISBN 0-669-00443-X

Copyright © 1981 by D.C. Heath and Company

All rights reserved. No part of this publication may be reproduced or transmitted in any form or by any means, electronic or mechanical, including photocopy, recording, or any information storage or retrieval system, without permission in writing from the publisher.

Published simultaneously in Canada

Printed in the United States of America

International Standard Book Number: 0-669-0044-X

Library of Congress Catalog Card Number: 75-39314

Contents

List of Figures		vii
List of Tables		ix
Preface and Acknowledgments		xi
Part I	Network Structure	1
Chapter 1	**Graph Theory Definitions and Properties**	3
	1-1 Graphs and Networks	3
	1-2 Routes, Components, and Cut-sets	4
	1-3 Cycles and Trees	6
	1-4 Connection Matrix	6
	1-5 Weights in Graphs	7
Chapter 2	**Trees**	9
	2-1 Introduction	9
	2-2 Minimum Spanning Trees	9
	2-3 Segments	12
	2-4 Trees with Constraints	12
	2-5 The Esau-Williams Algorithm	14
	2-6 Other Methods for Finding the Optimal Constrained Tree	17
	2-7 Practical Aspects	18
Chapter 3	**Routes**	21
	3-1 Introduction	21
	3-2 Weights	21
	3-3 Shortest Routes for All Node Pairs	22
	3-4 Shortest Routes from a Specified Node to All Other Nodes	27
	3-5 Generating Longer Routes	29
	3-6 Node-disjoint and Link-disjoint Routes	31
Chapter 4	**Reliability of Networks**	35
	4-1 Introduction	35
	4-2 Cohesion and Connectivity	35
	4-3 Finding the Cohesion of an Undirected Graph	36

	4–4 Finding the Connectivity of an Undirected Graph	41
	4–5 Probabilistic Link and Node Failures	44
Part II	**Switched Networks**	**51**
Chapter 5	**Circuit-Switched Networks**	**53**
	5–1 Introduction	53
	5–2 The Full-availability Trunk Group	54
	5–3 The Trunk Group under Other Assumptions	60
	5–4 Alternate Routing	64
	5–5 Blocking Probabilities in Networks	74
Chapter 6	**Message-Switched Networks**	**85**
	6–1 Introduction	85
	6–2 The Single-server Queue	86
	6–3 Kleinrock's Model	89
	6–4 Network Optimization: The Capacity Assignment Problem	92
	6–5 Network Optimization: The Flow Assignment Problem	98
Chapter 7	**Packet-Switched Networks**	**103**
	7–1 Introduction	103
	7–2 Extensions of Kleinrock's Model	104
	7–3 Network Optimization	108
Chapter 8	**Centralized Computer Networks**	**113**
	8–1 Introduction	113
	8–2 Estimation of Response Time	113
	8–3 Selecting Locations for Multiplexers/Concentrators	121
Chapter 9	**Least-cost-route Selection**	**129**
	9–1 Introduction	129
	9–2 Problem Statement	130
	9–3 The Full-access Trunk Group with Blocking	133
	9–4 The Full-access Trunk Group with Delays	136
	Index	**141**
	About the Author	**145**

List of Figures

1–1	Graph	4
1–2	Subgraph of the Graph of Figure 1–1	4
1–3	Graph with Two Components	5
1–4	Tree	7
1–5	Graph with Link Weights	8
2–1	(*a*) Graph (*b*) Minimum Spanning Tree	11
2–2	Network Layout, Showing Three Segments	13
2–3	Alternative Tree for Figure 2–2	13
2–4	Example for Esau-Williams Algorithm	15
2–5	Solution Tree for Figure 2–4	16
2–6	MST for Figure 2–4	17
3–1	FORTRAN Form of Floyd Algorithm, Steps 2 through 5	23
3–2	Example of Floyd Algorithm	25
3–3	Graph in which Only One Link Ends at Note x	26
3–4	Example of Node-Disjoint Routes	32
4–1	Directed Graph	36
4–2	Directed Graph with Feasible Flow Pattern	38
4–3	Example of Labeling and Augmentation Algorithm	40
4–4	(*a*) Original Undirected Graph $G(N, L)$ (*b*) Directed Graph $G * (N, L*)$	42
4–5	Graph $\hat{G}$ Corresponding to Figure 4–4(*b*)	43
4–6	Network to Illustrate Kleitman's Theorem	45
4–7	Network for Terminal Reliability Calculation	46
5–1	Full versus Partial Availability	61
5–2	Network to Illustrate Alternate Routing	65
5–3	Number of Circuits Busy versus Time for a Trunk Group of 14 Circuits	66

5–4	Trunk Groups 1 through n Overflowing to One Alternate Route	69
5–5	Network with Alternate Routing	75
5–6	Fragment of Network	79
6–1	Message-switched Network	86
6–2	A Unidirectional Link as a Single-server Queue	87
6–3	Three-node Network	91
6–4	Network for Optimization Example	93
8–1	(a) Nonhierarchical Network (b) Hierarchical Network	114
8–2	Initialized Network for Drop Algorithm	124
8–3	A Minimum-cost Solution for the Network of Figure 8–2 with One Concentrator	125
9–1	A "Black Box" for Least-cost-route Selection	130
9–2	Monthly Cost versus Usage for One Trunk, Two Different Types	131
9–3	Carried Traffic or Load per Trunk with $c = 5$ Trunks and $a = 3.0$ Erlangs	136

List of Tables

5–1	Erlang B Capacity Table	57
5–2	Maximum Peakedness Ratio versus Number of Circuits	68
5–3	Number of Circuits for a Blocking Probability $p = .01$ for Overflow Traffic	70
5–4	Example of Calculation of Weighted Mean Peakedness Ratio	73
5–5	Node-pair Blocking Probabilities for Network of Figure 5–5	83
6–1	Arrival Rates, Messages per Second	94

Preface and Acknowledgments

My purpose in writing this book was to assemble in one place a number of definitions, formulas, and algorithms that I have found useful in analyzing communications networks and that I hope will be useful to others in the field. In many instances, a network may be synthesized by repeatedly applying an analytic method and modifying the input data until a desired result is achieved. Also some of the procedures are directly applicable to synthesis, which extends the scope a little beyond what is implied by the title. The book is directed primarily to communications systems engineers, computer programmers, and students who need to analyze or design practical networks for telephone, data transmission, and other modes of communication. To the extent that the book points out the limitations of existing techniques, it may be helpful to research workers.

The book consists principally of descriptions of procedures that can be used to achieve particular results. I have tried to state clearly the meanings of all terms that are used; the required input data; the precise problem that is solved and the form of the results; and the limitations on the validity of the solutions. For brevity, there are no proofs given. Since many procedures now widely used in the field are partly or entirely heuristic, I have taken care to distinguish between proven solutions, in the mathematical sense, and others. Where possible, references for each procedure are given in which the reader may find proofs in many cases or other substantiation. A note on style: In the algorithms, the notation "$a \leftarrow b$" means that b is substituted for a, or that the value of b replaces the value of a.

Part I deals with the structural properties of networks. Since many useful procedures have come from graph theory, chapter 1 contains a selection of the most important definitions and graph properties. The subject of trees, which is important in several later applications, for example, multidrop arrangements, is treated in chapter 2. In chapter 3 routes are treated from the structural point of view, that is, apart from specific methods of alternate routing in networks. The concern here is with procedures for finding routes, especially when constraints are imposed.

Reliability in networks has received much attention, especially in military systems, where the vulnerability of nodes and branches (switching centers and communications links) is a concern. In order to simplify the treatment of this subject, I discuss in chapter 4 those aspects of reliability (or availability, or survivability) that are purely structural, that is, that can be analyzed from the graph properties of networks. The common measures of reliability relate to the proportion of node pairs that remain in communication when certain nodes or branches are destroyed or otherwise removed from service, or, in the next stage of complexity, when these things happen with specified probabilites. As there does not appear to be general agreement in the field on a set of reliability measures, I give a number of procedures that have been useful in practice; these embrace most of the principles that are required for estimating further reliability characteristics.

Part II introduces switching in networks. Broadly construed, the term means any method for establishing node-to-node connections that are not permanent. Circuit switching, also known as "line switching," is the basis of many telephone and other systems; connections are made to establish paths between nodes on demand and are then broken to make the branch capacities available to other network users. The treatment in chapter 5 begins with some fundamentals of traffic and congestion theory and proceeds to alternate routing and analytic models of networks.

Chapter 6 deals with message switching, also called "store-and-forward" switching, in which the network receives a message at a node and, rather than transmitting it directly to another node over a temporary path between the two, stores it and transmits it later when a suitable path is available; the storage and retransmission may be repeated at intermediate nodes. At the cost of complexity of the switching equipment, such networks achieve a higher utilization of the branches, relative to their capacities, than is generally possible with circuit switching. The principle model for message-switched networks, that of Kleinrock, is explained, with the necessary preliminary information borrowed from the theory of single-server queues or waiting lines.

Packet switching, treated in chapter 7, is relatively new to the field, although the technique was considered in 1964 by Baran. In some respects it is similar to message-switching, but the "messages" are packets, or blocks, of data that compose individual node-to-node messages. Much of the analysis of packet-switched systems has been by means of simulation, which is beyond the scope of this book. In chapter 7, I cite those analytic methods and algorithms that have been developed for this class of switching networks.

In chapter 8 I apply some of the results of chapter 2, along with the single-server queue model that underlies chapters 6 and 7, to the analysis of centralized computer networks. The problems of greatest practical interest are the estimation of response time and the selection of locations for multiplexers, concentrators, or other access nodes.

Least-cost-route or -trunk selection is a matter of current interest in private switching systems such as private branch exchanges (PBXs). Chapter 9 examines the principal analysis methods for least-cost routing systems in which calls may be either blocked or delayed.

Acknowledgments

I appreciate the financial support that I have had from my employer, Arthur D. Little, Inc., and in particular the help of Dr. Charles J. Kensler and Alan B. Kamman in arranging for this support. In addition, I have had the benefit of reviews of a substantial part of the manuscript by two coworkers, Dr. Donald B. Rosenfield and Dr. V Kevin M. Whitney.

**Part I
Network Structure**

1 Graph Theory Definitions and Properties

1-1 Graphs and Networks

A communications network consists of transmission paths that connect locations such as switching centers, concentrators, or terminals. We will model a network by a *graph*, which is a representation of the network that is independent of the makeup of the paths, for example, radio or cable, and of the functions accomplished at the locations that the paths connect, for example, switching, concentration, or information entry or receipt. Thus a graph consists of a finite set of points called *nodes*, connected by lines called *links*; the nodes correspond to the switching points, concentrators, terminals, and so on, and the links correspond to the transmission paths. In other discussions of graph theory, the term *vertex* is often used for a node, and terms such as *edge*, *arc*, or *branch* are used for a link.

An example of a graph is shown in figure 1–1. The nodes are the small circles with numbers in them, and the links are the lines that connect them. This description is adequate to picture the graph, but a proper graph theory definition is that the graph is a finite set of nodes (nodes 1, 2, 3, 4, and 5 in figure 1–1) and a set of node pairs [pairs (1, 2), (2, 3), (2, 4), (1, 4), (3, 4), and (3, 5) in figure 1–1], the latter set describing the links. In general, if the set of nodes is N and the set of links is L, we will refer to the graph as $G(N, L)$. A *subgraph* of $G(N, L)$ is a graph with a set of nodes that is a subset of N and set of links that is an appropriate subset of L. Figure 1–2 shows a subgraph of figure 1–1, obtained by removing node 2 and links (1, 2), (2, 3), and (2, 4). If $G_1(N_1, L_1)$ is a graph whose set of nodes is N_1 and whose set of links is L_1, and similarly for $G_2(N_2, L_2)$, then the *union* $G_1 \cup G_2$ of G_1 and G_2 is the graph whose set of nodes is $N_1 \cup N_2$ and whose set of links is $L_1 \cup L_2$.

The graph of figure 1–1 has certain properties in common with all graphs that we will discuss in this book unless a specific exception is necessary. It is *undirected*, that is, each link is arranged for communication in either direction between the nodes that it connects. Thus we will not distinguish between link (i, j) and link (j, i), i and j being node numbers. We will refer to nodes i and j as the *ends* of the link, and to the number of links ending at a node as the *degree* of the node; in figure 1–1, the degree of node 2 is 3. The graph has no link (i, j), both of whose ends are the same. Finally, any pair of nodes is connected by at most one link; of course, there may be no link, as in the case of nodes 1 and 5 of figure 1–1. Many important applications of graph

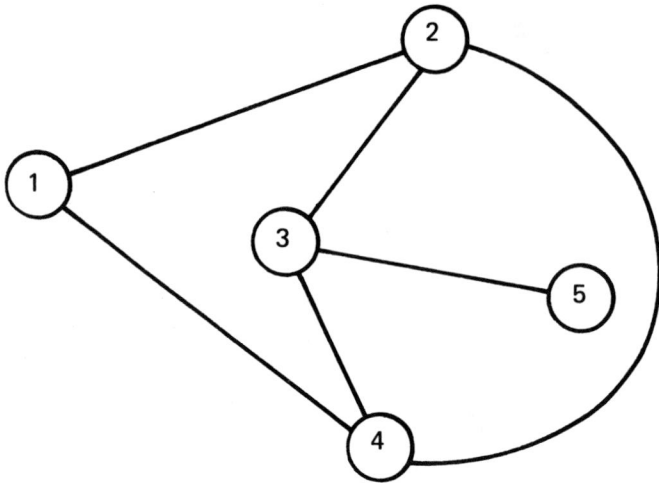

Figure 1-1. Graph

theory are based on graphs that lack one or more of these simplifying properties [1, 2]. As we will see, these properties will apply in most cases of interest here.

1-2 Routes, Components, and Cut-sets

If i and j are distinct nodes in $G(N, L)$, a *route from i to j* is a subset of L whose members can be arranged in an ordered list $(i, i_1), (i_1, i_2), \ldots, (i_k, j)$ with the following properties: (a) i appears only as one end of the first link,

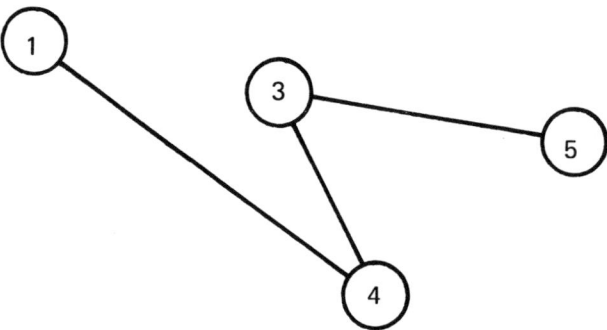

Figure 1-2. Subgraph of the Graph of Figure 1-1

Graph Theory Definitions and Properties 5

and j appears only as one end of the last link; and (b) if $k \geq 1$, each other node $i_1, i_2, \ldots, i_k$ appears exactly twice, as the common end of some link in the list and the link immediately following it. The reader should not be alarmed by this graph theory definition of a route in an undirected graph; it agrees with the commonsense notion of a communications route, properties a and b assuring that the route does not contain loops or retracings of a link. Any route from i to j corresponds in a unique and obvious fashion to a route from j to i.

A graph is *connected* if, for any distinct nodes i and j, there is a route from i to j. The graphs of figure 1-1 and figure 1-2 are connected. If a graph is not connected, it is *disconnected*. The graph of figure 1-3 is disconnected since it has no route from, for example, node 1 to node 4. If G' is a connected subgraph of graph G such that no other subgraph that contains G' is connected—that is, G' is a maximal connected subgraph [1]—then G' is called a *component* of G. In figure 1-1 the subgraph consisting of nodes 1, 2, and 3, and links (1, 2) and (2, 3), is one of the two components.

If L' is a subset of the links of a connected graph G such that removing the links in L' results in a graph with two or more components, and if further no subset of L' other than L' itself has this property—that is, L' is a minimal set of links with respect to this property [1]—then L' is called a *link cut-set of G*. In figure 1-1 the set of links consisting of (2, 3) and (3, 4) is a link cut-set. Frank and Frisch show that removing a link cut-set from a connected graph results in a graph with *exactly* two components [1].

If N' is a subset of the nodes of a connected graph G such that removing the nodes in N' results in a graph with two or more components, and if further no subset of N' other than N' itself has this property, then N' is called a *node cut-set of G*. In figure 1-1, the set consisting of node 3 is a node cut-set; one

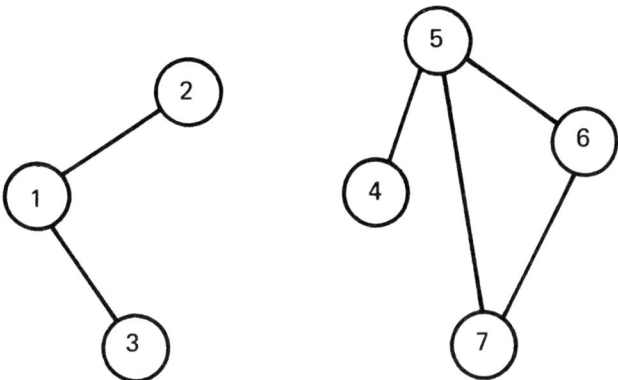

Figure 1-3. Graph with Two Components

of the components that results from its removal is the subgraph consisting of the solitary node 5.

1-3 Cycles and Trees

A *cycle* in a graph has a definition that is much like that of a route, section 1-2, but we require that the starting and ending nodes be the same. Thus a cycle is a subset of L whose members can be arranged in an ordered list $(i, i_1), (i_1, i_2), \ldots, (i_k, i)$ with the properties: (a) i appears only as one end of the first link and one end of the last link; and (b) each other node appears exactly twice, as the common end of some link in the list and the link immediately following it. Since our definition of a graph, section 1-1, does not permit a link of the form (i, i), we must have $k \geq 1$; in fact, since a link cannot be repeated in a subset of L [recall that link (i,j) is the same as link (j,i)], we have $k > 1$. A cycle is just a loop such as (5, 6), (6, 7), (7, 5) in figure 1-3.

A *tree* is a connected graph that contains no cycles, as illustrated in figure 1-4. Some interesting properties of a tree are: (a) removal of any link disconnects the graph; (b) there is exactly one route from any node to any other node; and (c) if n is the number of nodes, the graph contains n-1 links [3]. We will refer to any tree whose set of nodes is N as a *tree on N*.

1-4 Connection Matrix

If graph $G(N, L)$ contains n nodes, it can be completely described by the n-by-n matrix A whose i, jth element—that is, the element in the ith row and the jth column—$a_{ij} = 1$ if there is a link (i, j) or $= 0$ if there is not such a link. For the graph of figure 1-1,

$$A = \begin{bmatrix} 0 & 1 & 0 & 1 & 0 \\ 1 & 0 & 1 & 1 & 0 \\ 0 & 1 & 0 & 1 & 1 \\ 1 & 1 & 1 & 0 & 0 \\ 0 & 0 & 1 & 0 & 0 \end{bmatrix}$$

This is the *connection* matrix.

For the undirected graphs with which we will deal, A is symmetric, that is, $a_{ij} = a_{ji}$.

Graph Theory Definitions and Properties

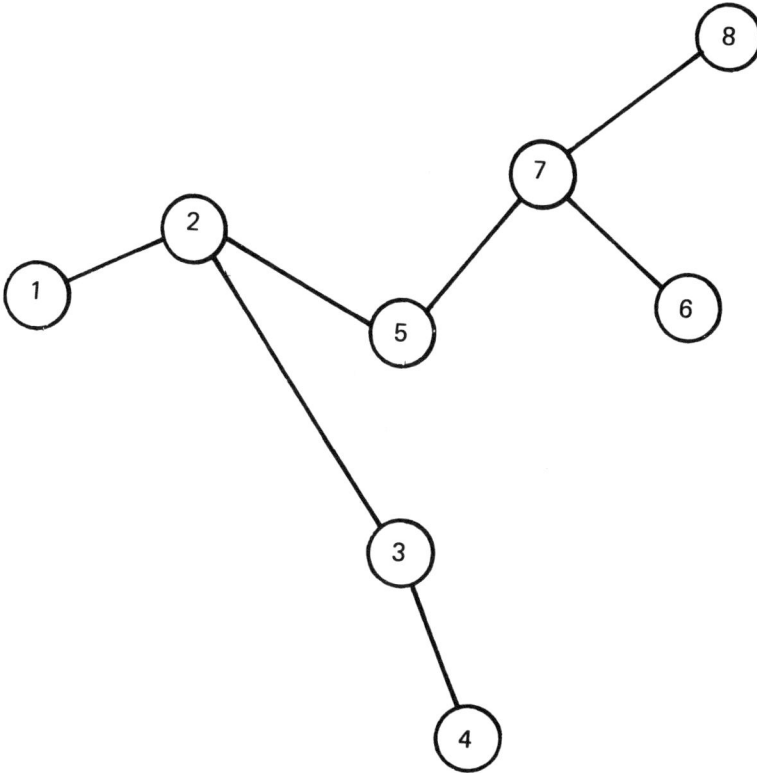

Figure 1-4. Tree

1-5 Weights in Graphs

It is often convenient to assign each link of a graph a real number called its *weight*. For example, the weight may be the length of the link, its cost, its transmission capacity, or other useful quantity. In any analysis, each link may have several different kinds of weights, denoted by different symbols. Weights are shown as numbers next to the links, as in figure 1-5; link (2, 4), has weight 3. Since we make no distinction between link (i,j) and link (j,i), neither will we distinguish between the weight of link (i,j) and that of link (j,i). Thus if w_{ij} is the weight of link (i,j), we will have $w_{ij} = w_{ji}$. We will adopt the convention $w_{ii} = 0$.

The weight of a cycle or route is the sum of the weights of its links. For

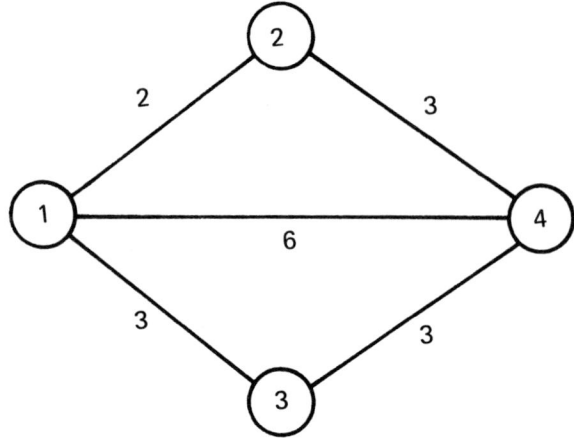

Figure 1-5. Graph with Link Weights

example, in figure 1-5, the route (1, 2), (2, 4) from node 1 to node 4 has weight 5, and the route (1, 4) has weight 6.

If we assign to each node a real number, we have a set of node weights, which may represent cost, switching capacity, and so forth.

In any discussion of a graph, such as a tree, in which it is clear that only link weights are involved, we will use terms such as the "weight of the tree" to mean the sum of the link weights. Where it is necessary, we will distinguish in a graph between the link weights, node weights, and various quantities calculated from them.

References

1. Frank, H., and Frisch, I.T. *Communication, Transmission, and Transportation Networks*. Reading, Mass.: Addison-Wesley, 1971.
2. Whitehouse, G.E. *Systems Analysis and Design Using Network Techniques*. Englewood Cliffs, N.J.: Prentice-Hall, 1973.
3. Knuth, D.E. *The Art of Computer Programming. v. 1, Fundamental Algorithms*. Reading, Mass.: Addison-Wesley, 1968.

2 Trees

2–1 Introduction

We recall from section 1–3 that a tree is a connected graph that contains no cycles. In the design of networks for the connection of terminals to computers, the following problem arises: given a fixed set of nodes such as the cities in which the terminals and computer are located, find a tree on these nodes that minimizes total link (communcations line) costs, perhaps subject to some constraints. The cost of a link of a particular capacity is often related to its length, and a typical constraint is that the nodes must be connected so that the traffic they generate does not overload the links. In microwave route layout, a minimum-length tree, with the terminal stations as nodes, is often useful as a starting point.

Since the number of trees on n nodes in n^{n-2}, it is impractical, except for small n, to generate all possible trees and find the cost of each as well as test it for compliance with any constraints. We therefore describe first a basic procedure for finding a *minimum spanning tree* (MST), that is, one that has minimun weight but is not subject to any constraints. Following this, we take up procedures for finding trees with minimum-weight properties when there are constraints.

2–2 Minimum Spanning Trees

We start with a set of nodes, N, whose members are at the outset not connected to each other. The object is to find a tree on N that has minimum weight among all possible trees.

It is convenient to imagine a graph, G, whose set of nodes in N, in which there is a link (i,j) for each node pair i,j ($i \neq j$); G is said to be *completely connected*. Each link in G has a weight, w_{ij}. In most practical cases, we will have $w_{ij} \geq 0$, but the solution that we give for finding the minimum spanning tree applies as well when some or all of the weights are negative. If it is impossible or undesirable to have a link between two particular nodes i and j, we simply assign w_{ij} the value infinity, or, for computer usage, some large positive value that can be recognized in a program.

The basic procedure is given by Prim [1]. Let us denote the desired minimum spanning tree (MST) by T. We start with an arbitrary node, say node 1, and construct a two-node subgraph of T by connecting it to the node, say j^*, that minimizes w_{1j}. (If, for example, w_{ij} is the link length, then j^* is the nearest node to node 1.) Let us call the set of nodes in this subgraph of T, consisting of nodes 1 and j^*, N_2. We now construct a three-node subgraph of T by connecting one of the nodes in N_2, say, m, to another node k, not in N_2, where nodes m and k are chosen so that

$$w_{mk} = \min(w_{mr} : m \in N_2, r \in N - N_2)$$

in which $N - N_2$, in the customary set-theory notation, means the set of nodes that belong to N but not to N_2. That is, of all possible links from any node in N_2 to any node in $N - N_2$, the added link has minimum weight. We complete T by adding links in this way: at each step, the added link has minimum weight of all possible links from any node among those so far connected to any node among those not yet connected.

An example is shown in figure 2–1. Figure 2–1(a) shows a set of five nodes and weights w_{ij}; we may assume that omitted links such as (1, 5) have infinite weight. If we start with node 1, for example, the steps are as follows:

Nodes in Partial MST	Link Added
1	(1, 2)
1, 2	(2, 3)
1, 2, 3	(3, 5)
1, 2, 3, 5	(2, 4)
1, 2, 3, 4, 5	–

The resulting MST is shown in figure 2–1(b), and has weight 8.

A formal statement of this algorithm follows [2]. In this adaptation, the possiblity that no minimum spanning tree exists, that is, that the hypothetical graph G is disconnected, is eliminated since this case is not of practical interest in the customary applications of the algorithm to communications networks.

Prim Algorithm (Minimum Spanning Tree)

Given a set of n nodes N, and a set of link weights $w_{ij}(w_{ij} = w_{ji}, w_{ii} = 0)$, find a minimum spanning tree on N. The MST is identified by its connection matrix, A. The algorithm yields a unique MST if the link weights, other than those that may be infinite, are distinct.

Step 1 [Initialization]. Define N_k as a subset of N containing k nodes; set $k \leftarrow 1$, and $N_1 \leftarrow \{1\}$. Define t_j as a set of test values and set $t_j \leftarrow w_{1j}$,

Trees

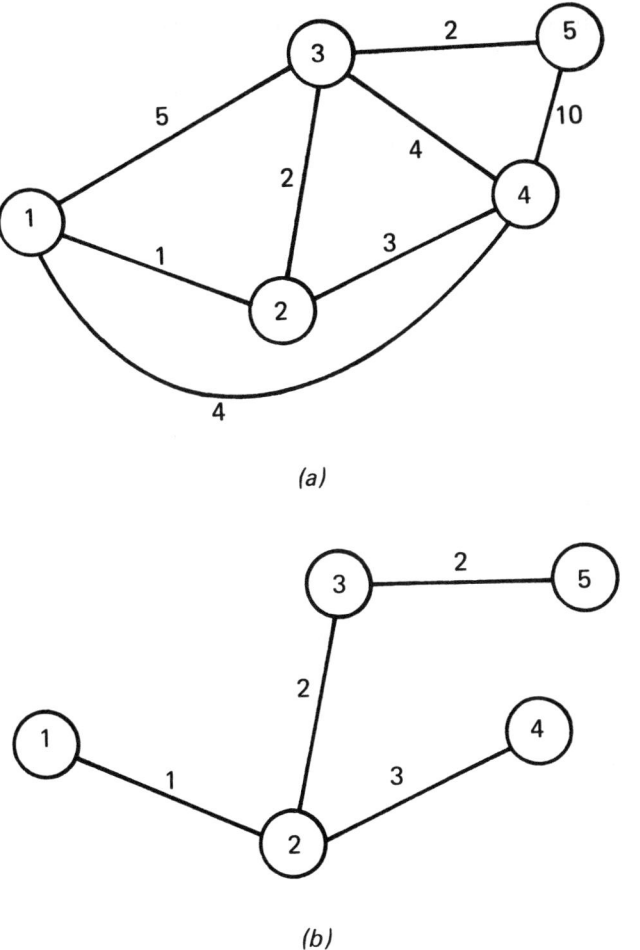

Figure 2–1. (a) Graph (b) Minimum Spanning Tree

$j = 1, 2, \ldots, n$. Set $a_{ij} \leftarrow 0, i, j = 1, 2, \ldots, n$. (The arbitrary node with which we start is node 1.)

Step 2 [Add one more node and modify A correspondingly]. Find node j^* such that $t_{j^*} = \min \ (t_i : i \in N - N_k)$. Find node $i^* \in N_k$ for which $w_{i^*j^*} = t_{j^*}$. Set $a_{j^*j^*} = a_{j^*i^*} \leftarrow 1$. If $k + 1 = n$, stop. Otherwise, set $k \leftarrow k + 1$ and $N_k \leftarrow N_{k-1} \cup \{j^*\}$.

Step 3 [Update test values t_j]. For $j \in N - N_k$, set $t_j \leftarrow \min(t_j, w_{j^*j})$. Go to step 2.

Prim's algorithm as given above is somewhat inefficient in its use of computer time, especially for sparse networks, that is, those in which relatively few of the $n(n - 1)/2$ possible links are usable. Its ease of implementation and modest storage requirements make it practical to use in most communications network problems, as compared to more complex and more efficient versions or to alternative algorithms [2].

2-3 Segments

In the MST of figure 2-1(b), no node is distinguished from any other. Suppose, though, that there is a computer center at node 2, and terminals at nodes 1, 3, 4, and 5. A typical network layout corresponding to the MST is then shown in figure 2-2. It consists of three subnetworks: (a) the computer center, terminal node 1, and a two-point communications line connecting them; (b) the computer center, terminal node 4, and a two-point communications line connecting them; and (c) the computer center, terminal nodes 3 and 5, and a three-point communications line connecting them. We will refer to the graphs of each of these subnetworks as *segments* [3]. A segment is a connected subgraph of a tree (and is thus a tree itself) that contains exactly one link that ends at a specific node, k. In figure 2-2, for example, $k = 2$.

For any tree, there is a unique set of segments, relative to some node, k, such that each node of the tree, other than k, appears in exactly one segment.

2-4 Trees with Constraints

Referring again to figure 2-2, the links (1, 2), (2, 3), and so on, may have capacities high enough to handle all the communications for each segment. For example, link (2, 3) may have enough capacity for all communications between terminal node 3 and the computer center, as well as those between terminal node 5 and the computer center, in both directions. (We do not preclude communications between terminal nodes. It is convenient for this discussion to assume that such communications take place via the computer center, even for terminal nodes in the same segment.) If this condition is met for all segments, then the MST is the minimum weight tree for the given communications requirements, and its segments, relative to the computer-center node, constitute a practical network layout.

On the other hand, there may be constraints that require a solution other than the MST. Suppose, for example, that the communications traffic load of terminal node 5 in figure 2-2 is too great for link (3, 5); then the segment containing nodes 2, 3, and 5 is not feasible. Provided that the traffic loads of nodes 5 and 4 are not too great relative to the capacities of links (2, 4) and

Trees

Computer center: node 2

Terminals: nodes 1, 3, 4, 5

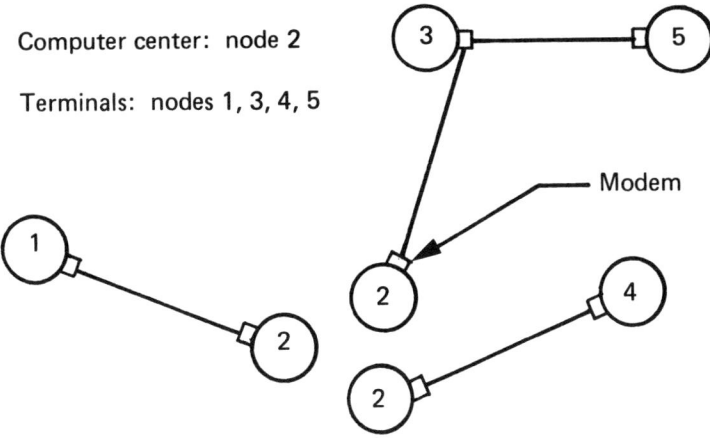

Figure 2–2. Network Layout, Showing Three Segments

(4, 5), we may connect node 5 to node 4, deleting link (3, 5) and adding link (4, 5). The new tree that results is shown in figure 2–3; it has weight 16, as compared to the MST weight of 8. This illustrates the principle that the MST is a lower bound for solutions of the constrained problem.

We will illustrate the solution of the minimum-weight tree problem with constraints by means of two particular constraints. The first is that each terminal node i has a *message rate*, $m_i \geq 0$, and the sum of the message rates for the terminal nodes in any segment, relative to a specified node, k, may not exceed a fixed maximum, M. (In chapter 6 we will take up the question of how M is established. The message rate is a measure of terminal node activity, for example, the number of messages of known average duration per

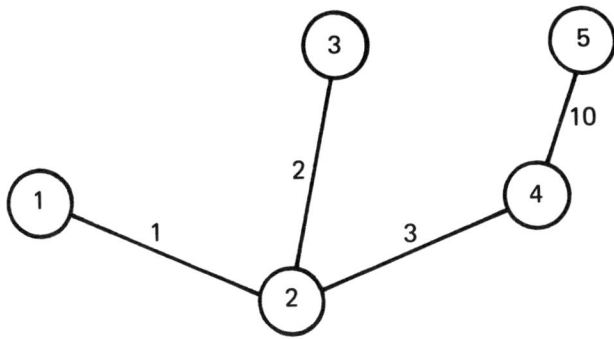

Figure 2–3. Alternative Tree for Figure 2–2

unit time.) To assure that there is a feasible solution, we assume $m_i < M$ for all terminal nodes i.

The second kind of illustrative constraint is that no segment may contain more than K terminal nodes. If $K = 1$, the only feasible tree is the "star," that is, the tree in which each terminal node is connected by its own link to the computer center. We will thus assume $K \geq 2$.

2-5 The Esau-Williams Algorithm

A very useful algorithm is that of Esau and Williams [4]. Unlike the Prim algorithm (section 2-2), it is heuristic; it does not yield the minimum-weight tree in all cases, but it is quite satisfactory in practice. The following presentation is adapted from Kershenbaum and Chou [5].

Esau-Williams Algorithm (Minimum-weight Tree with Constraints)

Given a set of n nodes, N, node 1 being the computer-center and nodes 2, 3, ..., n the terminal nodes, with link weights w_{ij} ($w_{ij} = w_{ji}$, $w_{ii} = 0$, $i,j = 1, 2, ..., n$), find the minimum-weight tree on N subject to the following constraints: (a) the sum of the message rates m_i of the terminal nodes in each segment relative to node 1 may not exceed a fixed maximum, M; and (b) there may not be more than K terminal nodes in a segment. (These constraints are subject to the conditions given in section 2-4.) The tree is identified by its connection matrix A.

Step 1 [Initialization]. Define d_i, $i = 2, 3, ..., n$ as a set of terminal node weights; set $d_i \leftarrow w_{i1}$. For the completely connected graph G whose set of nodes is N, define C_i as a component containing node i; set $C_i \leftarrow \{i\}$, $i = 1, 2, ..., n$. Define t_{ij}, $i,j = 1, 2, ..., n$ as a set of test values, and set $t_{ij} \leftarrow w_{ij} - d_i = w_{ij} - w_{i1}$; set $t_{ij} \leftarrow \infty$ if there is no feasible link between i and j ($w_{ij} = \infty$), or if $m_i + m_j > M$. Set $a_{ij} \leftarrow 0$, $i, j = 1, 2, ..., n$.

Step 2 [Search for possible new link]. Find

$$t_{i^*j^*} = \min\ (t_{ij} : i,j \text{ not in the same component})$$

If $t_{i^*j^*} = \infty$, stop.

Step 3 [Test permissibility of adding link (i^*,j^*) against constraints]. For convenience, denote by C' the new component, $C_{i^*} \cup C_{j^*}$, that would be formed by adding link (i^*, j^*). If C' does not contain node 1, then the constraints are met if and only if (a) $\Sigma (m_i : i \in C') \leq M$; and (b) the number of terminal nodes in C' is less than or equal to K. If C' contains node 1, then the constraints are met if and only if (c) the sum of the message rates for the terminal nodes in any segment changed or created by forming C' is less than or equal to M; and (d) the number of terminal nodes in any such segment is less than or equal to K. In either case, if the constraints are not met, set $t_{i^*j^*} \leftarrow \infty$ and go to step 2. (The segments are relative to node 1.)

Step 4 [Add link (i^*, j^*), update the node weights and the t_{ij}, and form new component C']. Set $a_{i^*j^*} = a_{j^*j^*} \leftarrow 1$. For $i \in C_{j^*}$, set $d_i \leftarrow d_{j^*}$, and then update the t_{ij}, that is, for $i \in C_{i^*}$, set $t_{ij} \leftarrow w_{ij} - d_i$. Set $C_{i^*} = C_{j^*} \leftarrow C'$. Go to step 2. (It is sufficient in this algorithm to keep track solely of the nodes in each C_i; connectivity within each component is available, if necessary, in A.)

The working of the algorithm is made clear by considering an example such as the graph shown in figure 2–4 with the constraints $M = 50$ and $K = 3$. At the first pass through step 2, we find that $t_{i^*j^*} = t_{34} = w_{34} - d_{31} = 1 - 12 = -11$. As explained by Martin, it is helpful to imagine that we obtain the solution by starting with a hypothetical star tree, in which each terminal node is directly connected to node 1 by its own link, and then reducing the weight, relative to that of the star, by creating segments that contain two or more terminal nodes [6]. (Under the conditions on the constraints, the star is always a feasible solution, although it will not in general be a minimum-weight solution. The star is not introduced explicitly in

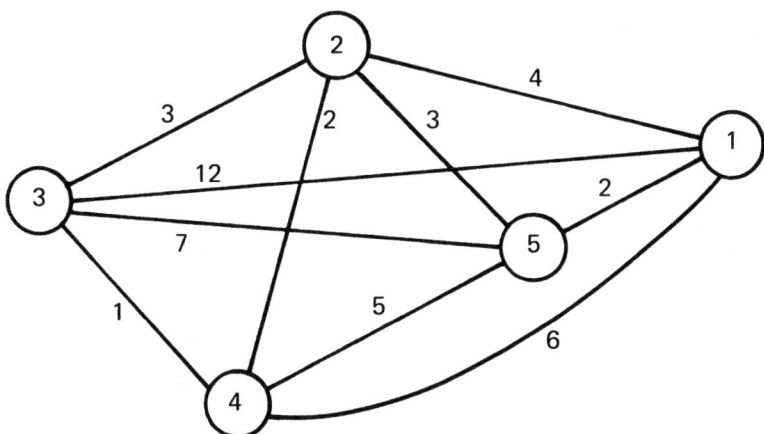

Figure 2–4. Example for Esau-Williams Algorithm
Terminal node message rates: $m_2 = 30$, $m_3 = 10$, $m_4 = 40$, $m_5 = 10$.

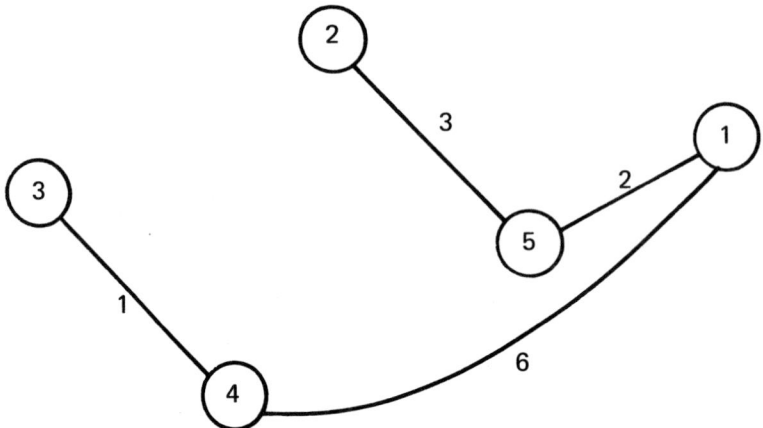

Figure 2-5. Solution Tree for Figure 2-4

the algorithm.) The t_{ij} are "trade-off costs," and the meaning of the minimal nature of $t_{3\,4}$ is that we achieve the greatest reduction in tree weight, relative to that of the star, by connecting node 3 to node 4 and deleting link (1, 3). As required at step 3, we must check the effects of adding the possible new link (3, 4) against the constraints. Since $m_3 + m_4 = 10 + 40$ is not more than $M = 50$ and since up to three terminal nodes per segment are allowed, the constraints are met, and link (3, 4) is the start of the solution tree. By pursuing the algorithm to its conclusion, we find this tree as shown in figure 2-5. It contains two segments, one with nodes, 2, 5, and 1 and the other with nodes 3, 4, and 1, and has weight 12. An alternate feasible arrangement, in which the segments contain nodes (3, 2, 5, 1) and (4, 1), has weight 14, whereas the MST (figure 2-6) has weight 8.

Kershenbaum and Chou show that the Esau-Williams algorithm, as given previously, is a special case of "unified" algorithm. The specialization is in the way the node weights d_i are initialized in step 1 and updated in step 4 [5] With appropriate alternative treatments of the d_i, the unified algorithm becomes equivalent to that of Prim (section 2-2); to that of Kruskal [7], another widely used algorithm for finding the MST; or to another algorithm, attributed to Vogel, for finding the constrained MST.

For the Esau-Williams algorithm, Kershenbaum and Chou point out that the computation time to update the t_{ij} at step 4 is reduced by the implicit definition of the t_{ij} in terms of the w_{ij}, which do not change, and the d_i, which change only for the nodes in C_{i*}. For other algorithms that are special cases of the "unified" algorithm, computation time is reduced, with little or no sacrifice of quality of solution, that is, with little or no departure from a near-minimum tree, by restricting the search for a minimum value of t_{ij}, for any i,

Trees

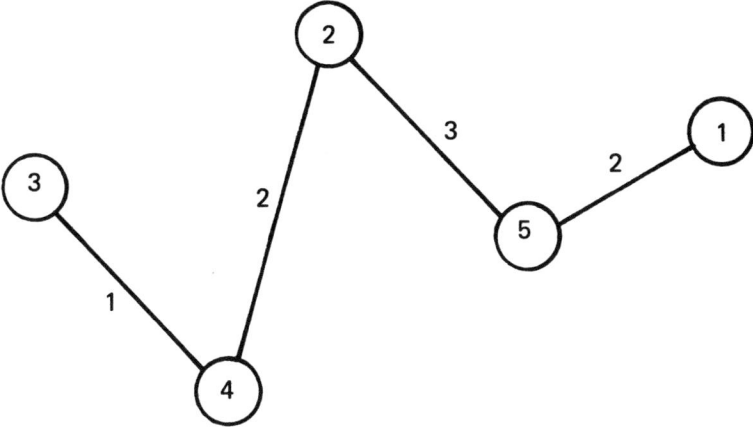

Figure 2-6. MST for Figure 2-4

to a small number of nearest neighbors of node i, that is, to a subset of N for which w_{ij} is less than some bound, say, w_{i1}. (It is generally not practical to connect a terminal node i to a node that is farther from i than is the computer, node 1, since we might as well connect i to 1 directly [5].)

If the constraints are not the same as the illustrative ones we have used, step 3 of the Esau-Williams algorithm must be modified accordingly. An example of a somewhat more general message-rate constraint is given by Karnaugh [8]: rather than the sum of the terminal node message rates for any segment being limited to a uniform maximum, M, the sum for a segment containing t terminal nodes is limited to M_t, with $M_K \leq M_{K-1} \leq \cdots \leq M_1$; as in section 2-4, K is the maximum allowable number of terminal nodes per segment. Our constraint is the specialization of Karnaugh's, with $M_K = M_{K-1} = \cdots = M_1 = M$.

2-6 Other Methods for Finding the Optimal Constrained Tree

As we have indicated in section 2-5, the Esau-Williams algorithm is by no means the only procedure for finding the minimum-weight tree with constraints. Indeed, there are several alternative heuristic procedures that may result in a total tree weight that is a little less than that obtained in an Esau-Williams solution. These include a variant of the "unified" algorithm of Kershenbaum and Chou [5]; an algorithm proposed by Elias and Ferguson [3] that exemplifies link-exchange procedures, in which a starting tree of low weight such as the MST is altered by successive replacements of one link by another in order to meet the constraints (see, for instance, the example in section 2–4); and the class of algorithms described by Karnaugh [8].

If it is important to pursue every possible avenue toward a minimum-weight solution the designer may wish to use one of these alternative algorithms. He should realize that the potential reduction in tree weight, using common weights such as airline distance or common carrier charges and numbers of nodes up to about 100, is only a few percent relative to Esau-Williams, whereas the alternative algorithms are in general more complex and tend to have greater execution times and storage requirements.

It is interesting to note that there is an exact solution, the branch-and-bound method [9], which is based on the following theorem: if in a minimum spanning tree terminal nodes $i_1, \ldots, i_r$ are directly connected to node 1—that is, there are links $(i_1, 1), \ldots, (i_r, 1)$ with $w_{i_k 1} < \infty$, $k = 1, \ldots, r$—the computer-center node, then there exists a minimum-weight constrained tree in which nodes $i_1, \ldots, i_r$ are directly connected to node 1. For example, in figure 2–6, the MST for figure 2–4, node 5 is directly connected to node 1, and the same is true in the constrained tree, figure 2–5; the latter, in this instance, is not only the Esau-Williams but also the optimal solution.

The branch-and-bound method, using figure 2–4 as an example, would start by partitioning all trees into two classes: (a) those in which link (5, 1) is present; and (b) those in which it is not. It is then necessary to search for an optimal constrained tree of the kind whose existence is assured by the preceding theorem only among members of class a. Within class a, the optimal solution cannot have lower weight than that of the MST; if the MST meets the constraints, it is the solution. If not, class a is further partitioned, and lower bounds are found on solutions within each subcalss, essentially by extensions of the theorem. The solution terminates when, at any iteration, a feasible solution is found whose weight is the lowest of the lower bounds for the subclasses being examined.

The branch-and-bound method is, in general, impractical since the computation time is substantially more than for the heuristic methods, for example, Esau-Williams, and tends to increase very rapidly with the number of nodes. Its principal application has been in testing heuristic methods; even in this role, the algorithm has not, within a practical number of iterations, obtained the optimal solution in all cases [3, 9].

2–7 Practical Aspects

The motivation for finding MSTs or minimum-weight constrained trees lies in the principle that common-carrier mileage charges for multipoint private lines are calculated for the tree that minimizes the charges to the user. Thus the MST for a multipoint line is the logical extension of the great circle or airline route for a two-point line, at least for terrestrial transmission paths. Two situations must be distinguished in assigning link weights. If per-mile charges are the same for all links, then w_{ij} equals the length (airline mileage) of link (i, j). On the other hand, if mileage charges must be calculated separately for each link, perhaps depending on link length, then w_{ij} equals the charges for link (i, j).

In addition to the mileage charges, there are also carrier charges for the termination of each link at each node; for local drops to terminals at the same node, that is, in the same city or exchange; and possibly for conditioning or special equalization of each link or of the entire multipoint line. These charges could be incorporated in the algorithms, but since they do not influence the selection of the tree, it is simpler to calculate them separately. This applies as well to the costs of modems, communications controllers, and other equipment at the nodes.

Just as a two-point line is rarely routed physically along a great circle path, a multipoint line will in general have a quite different physical layout from the tree, something that must be considered in the transmission engineering of the line [10, 11].

References

1. Prim, R.C. "Shortest Connection Networks and Some Generalizations," *Bell Syst. Tech. J.* 36:1389–1401, 1957.
2. Network Analysis Corporation. Fifth Semiannual Technical Report... for the Project, "Research in Store and Forward Computer Networks," AD–748338, June 1972.
3. Elias, D., and Ferguson, M.J. "Topological Design of Multipoint Teleprocessing Networks," *IEEE Trans. Communications* COM–22:1753–1762, November 1974.
4. Esau, L.R., and Williams, K.C. "On Teleprocessing System Design, Part II, A Method for Approximating the Optimal Networks." *IBM Syst. J.* 5(3):142–147, 1966.
5. Kershenbaum, A., and Chou, W. "A Unified Algorithm for Designing Multidrop Teleprocessing Networks." *IEEE Trans. Communications* COM–22(11):1762–1772, 1974.
6. Martin, J. *Systems Analysis for Data Transmission*. Englewood Cliffs, N.J.: Prentice-Hall, 1972.
7. Kruskal, J.B. "On the Shortest Spanning Subtree of a Graph and the Traveling Salesman Problem," *Proc. Amer. Math. Soc.* 7:48–50, 1956.
8. Karnaugh, M. "A New Class of Algorithms for Multipoint Network

Optimization." *IEEE Trans. Communications* COM–24(5):500–505, May 1976.
9. Chandy, K.M., and Russell, R.A. "The Design of Multipoint Linkanges in a Teleprocessing Tree Network." *IEEE Trans. Computers* C–21(10): 1062–1066, 1972.
10. Bell System Data Communications, Technical Reference. Data Communications Using Voiceband Private Line Channels. PUB41004, 1973.
11. Bell System Data Communications, Technical Reference. Multistation DATAPHONE® Digital Service. PUB41022 (Preliminary), 1974.

3 Routes

3-1 Introduction

The major practical problem relating to routes in communications networks is to find one or more routes between a specified pair of nodes, or between all pairs of nodes, subject to certain constraints. For example, we will see in part II that lists of such routes are required in the analysis of switched networks. Algorithms to generate routes are desirable because, somewhat as in the case of trees (see section 2–1), the number of routes, disregarding constraints, grows very rapidly with the number of ndoes, and it is thus impractical to enumerate all routes and select those that are acceptable. {For a completely connected graph with n nodes, the number of routes for any node pair approaches $e(n-2)!$ asymptotically with n, where $e = 2.718...$ [1].}

We take up first the constraint that has gotten the most attention in the literature, that of minimum weight; this is the shortest-route problem [2]. We then deal with an algorithm for generating routes with bounded (greater than minimum) weights, and finally with the problem of finding routes with weight constraints as well as constraints on common nodes or links.

3-2 Weights

In this chapter, a link weight is a nonnegative quantity such as the length or a cost that is a function of the length. A simple special case is that in which the weight of each link is 1; the weight of a route is then the number of links it contains.

In the most general route-finding problem, as it is approached in graph theory, it is necessary to eliminate the possibility that a route may contain a cycle of negative weight since, by traversing such a cycle repeatedly, the weight of the route can be made arbitrarily large in magnitude and negative. Here such a possibility is doubly ruled out from the start since we recognize only nonnegative link weights and since a route may not contain a cycle (see section 1–3). The latter restriction is of no moment in communications networks where there is no practical value to a route that contains a cycle.

In the remainder of this chapter, we will deal with a connected graph $G(N, L)$ having a set of n nodes, N, and a set of links, L. Each link (i, j) has a weight d_{ij}, with $d_{ij} = d_{ji} \geq 0$, and $d_{ii} = 0$; if there is no link (i, j), $d_{ij} = \infty$. To conform to general usage, we will consider "length" a synonym for "weight," and "shortest" a synonym for "minimum weight" in describing a route. However, the triangle inequality, $d_{ij} + d_{jk} \geq d_{ik}$ for distinct i, j, k does not necessarily hold. Further, in some applications of the techniques for generating minimum-weight routes, alternative kinds of weights are used. For example, if for each link, p_{ij} is the probability of the link's remaining operable in the face of possible destruction or malfunction, and we define a weight $w_{ij} = -\log p_{ij}$, then a minimum-weight route is one with the highest probability of being operable, assuming that this probability for a route is the product of the link probabilities. {Shek deals with graphs in which each link has not only a length d_{ij} but also a nonnegative cost, h_{ij} [3]. He shows how to find the shortest routes for which the route cost, defined as the sum of the link costs, does not exceed a specified bound.}

3-3 Shortest Routes for All Node Pairs

The algorithm of Floyd addresses most directly two questions: (a) What is the length of the shortest route between node i and node j, for all pairs i, j with $i \neq j$? and (b) what links make up the shortest route between i and j [4]? In the event that there are two or more shortest routes, the algorithm defines only one such route.

Floyd Algorithm
(Shortest-route Length Matrix and Successor Node
Matrix [5])

For graph $G(N, L)$, find S, the symmetric n-by-n matrix whose i,jth element s_{ij} equals the length of the shortest route between node i and node j, and R, the n-by-n matrix whose i,jth element $r_{ij}=$ the successor node of i in the shortest route between i and j. In the definition of a route between i and j, section 1-2, the successor node of i is the node following i in the list of links, viz., i_1. We may of course have $i_1 = j$. The way in which the R matrix is used to generate the shortest route is described later in the present section.

Step 1 [Initialize matrices]. For $i = 1, \ldots, n$ and $j = 1, \ldots, n$: set $s_{ij} \leftarrow d_{ij}$; $r_{ij} \leftarrow j$ if $d_{ij} < \infty$ and $i \neq j$; $r_{ij} \leftarrow 0$ if $d_{ij} = \infty$ and $i \neq j$, or if $i = j$.

Step 2 [Set value of intermediate node i]. For $i = 1, \ldots, n$ do steps 3 through 5.

Step 3 [Set j]. For $j = 1, \ldots, n$ and $j \neq i$, $s_{ji} < \infty$, do steps 4 and 5.

Routes

```
        DO 100 I = 1,ND
        DO 100 J = 1,ND
        IF (J .EQ.I.OR. S(J,I) .EQ. INF) TO TO 100
        DO 100 K = 1,ND
        IF (K .EQ. J .OR. K .EQ. I .OR. S(I,K) .EQ. INF) GO TO 100
        X = S(J,I) + S(I,K)
        IF (X .GE. S(J,K)) GO TO 100
        S(J,K) = X
        R(J,K) = I
100     CONTINUE
```

Figure 3-1. FORTRAN Form of Floyd Algorithm, Steps 2 through 5

$ND = n$, the number of nodes in G, and arrays S (real) and R (integer) are dimensioned (ND, ND). These arrays must be initialized, as in step 1 of the algorithm (see section 3-3), before this program is executed. *INF* is a large real number that substitutes for infinity.

Step 4 [Set k]. For $k = 1, \ldots, n$ and $k \neq j$, $k \neq i$, $s_{ik} < \infty$, do step 5.

Step 5 [Test for possibly shorter route from j to k via i, and update s_{jk} and r_{jk}]. If $s_{ji} + s_{ik} < s_{jk}$, set $s_{jk} \leftarrow s_{ji} + s_{ik}$ and $r_{jk} \leftarrow i$.

The FORTRAN program of figure 3-1 may help to clarify steps 2 through 5.

A useful interpretation of the algorithm is as follows [6]. We start with matrix $S^{(0)}$, with $s_{jk}^{(0)}$ equal to the length d_{jk}. Next, with $i = 1$, we find matrix $S^{(1)}$, in which $s_{jk}^{(1)}$ equals the length of the shortest route from j to k when node 1 (only) is allowed as a possible intermediate node. Continuing, matrix $S^{(m)}$ has $s_{jk}^{(m)}$ equal to the length of the shortest route from j to k when any subset of nodes $1, 2, \ldots, m$ may be intermediate nodes; the final result is $S = S^{(n)}$. At each iteration, that is, at each successive value of i in step 2,

$$s_{jk}^{(i)} = \min\,[s_{jk}^{(i-1)},\, s_{j,i}^{(i-1)} + s_{i,k}^{(i-1)}]$$

Similar remarks apply to matrix R.

In the example of figure 3-2, we have initially

$$S = \begin{bmatrix} 0 & 0.5 & 2.0 & 1.5 & \infty & \infty & \infty \\ 0.5 & 0 & \infty & \infty & 1.2 & 9.2 & \infty \\ 2.0 & \infty & 0 & \infty & 5.0 & \infty & 3.1 \\ 1.5 & \infty & \infty & 0 & \infty & \infty & 4.0 \\ \infty & 1.2 & 5.0 & \infty & 0 & 6.7 & \infty \\ \infty & 9.2 & \infty & \infty & 6.7 & 0 & 15.6 \\ \infty & \infty & 3.1 & 4.0 & \infty & 15.6 & 0 \end{bmatrix}$$

and

$$R = \begin{bmatrix} 0 & 2 & 3 & 4 & 0 & 0 & 0 \\ 1 & 0 & 0 & 0 & 5 & 6 & 0 \\ 1 & 0 & 0 & 0 & 5 & 0 & 7 \\ 1 & 0 & 0 & 0 & 0 & 0 & 7 \\ 0 & 2 & 3 & 0 & 0 & 6 & 0 \\ 0 & 2 & 0 & 0 & 5 & 0 & 7 \\ 0 & 0 & 3 & 4 & 0 & 6 & 0 \end{bmatrix}$$

After step 2 has been completed with $i = 2$, that is, after all routes in which nodes 1 or 2 may be intermediate nodes have been tested, we have

$$S = \begin{bmatrix} 0 & 0.5 & 2.0 & 1.5 & 1.7 & 9.7 & \infty \\ 0.5 & 0 & 2.5 & 2.0 & 1.2 & 9.2 & \infty \\ 2.0 & 2.5 & 0 & 3.5 & 3.7 & \infty & 3.1 \\ 1.5 & 2.0 & 3.5 & 0 & 3.2 & 11.2 & 4.0 \\ 1.7 & 1.2 & 3.7 & 3.2 & 0 & 6.7 & \infty \\ 9.7 & 9.2 & \infty & 11.2 & 6.7 & 0 & 15.6 \\ \infty & \infty & 3.1 & 4.0 & \infty & 15.6 & 0 \end{bmatrix}$$

and

$$R = \begin{bmatrix} 0 & 2 & 3 & 4 & 2 & 0 & 0 \\ 1 & 0 & 1 & 1 & 5 & 6 & 0 \\ 1 & 1 & 0 & 1 & 1 & 0 & 7 \\ 1 & 1 & 1 & 0 & 1 & 1 & 7 \\ 2 & 2 & 2 & 0 & 0 & 6 & 0 \\ 0 & 2 & 0 & 2 & 5 & 0 & 7 \\ 0 & 0 & 3 & 4 & 0 & 6 & 0 \end{bmatrix}$$

Notice here, for example, that $s_{46} = 11.2$, the length of route (4, 1), (1, 2), (2, 6); in the previous pass through step 2, with $i = 1$, the value of s_{42} was changed from its initial value of ∞ (nodes 4 and 2 not being connected by a link) to the value 2.0, the length of the route from 4 to 2 via 1.

Finally, at the conclusion of the algorithm, we find

$$S = \begin{bmatrix} 0 & 0.5 & 2.0 & 1.5 & 1.7 & 8.4 & 5.1 \\ 0.5 & 0 & 2.5 & 2.0 & 1.2 & 7.9 & 5.6 \\ 2.0 & 2.5 & 0 & 3.5 & 3.7 & 11.7 & 3.1 \\ 1.5 & 2.0 & 3.5 & 0 & 3.2 & 9.9 & 4.0 \\ 1.7 & 1.2 & 3.7 & 3.2 & 0 & 6.7 & 7.2 \\ 8.4 & 7.9 & 11.7 & 9.9 & 6.7 & 0 & 14.8 \\ 5.1 & 5.6 & 3.1 & 4.0 & 7.2 & 14.8 & 0 \end{bmatrix}$$

Routes

and

$$R = \begin{bmatrix} 0 & 2 & 3 & 4 & 2 & 2 & 3 \\ 1 & 0 & 1 & 1 & 5 & 5 & 1 \\ 1 & 1 & 0 & 1 & 1 & 5 & 7 \\ 1 & 1 & 1 & 0 & 1 & 1 & 7 \\ 2 & 2 & 2 & 2 & 0 & 6 & 2 \\ 5 & 5 & 5 & 5 & 5 & 0 & 5 \\ 3 & 3 & 3 & 4 & 4 & 3 & 0 \end{bmatrix}$$

The way in which we use the R matrix to generate the shortest route between a specified pair of nodes can be shown with an example from figure 3–2. If we wish to describe the shortest route between node 2 and node 7, we proceed from node 2 to the node indicated as $r_{2\,7}$ in the R matrix as it stands after the entire algorithm has been performed, that is, we go to node 1, and the first link in the desired route is (2, 1). Next we proceed to the node whose number is given as $r_{1\,7}$, the successor node to 1 in the shortest route from 1 to 7; this is 3, so we add link (1, 3). Finally, we see that the successor to 3 in the shortest route from 3 to 7 is 7, so the route is (2, 1), (1, 3), (3, 7), with length $s_{2\,7} = 5.6$.

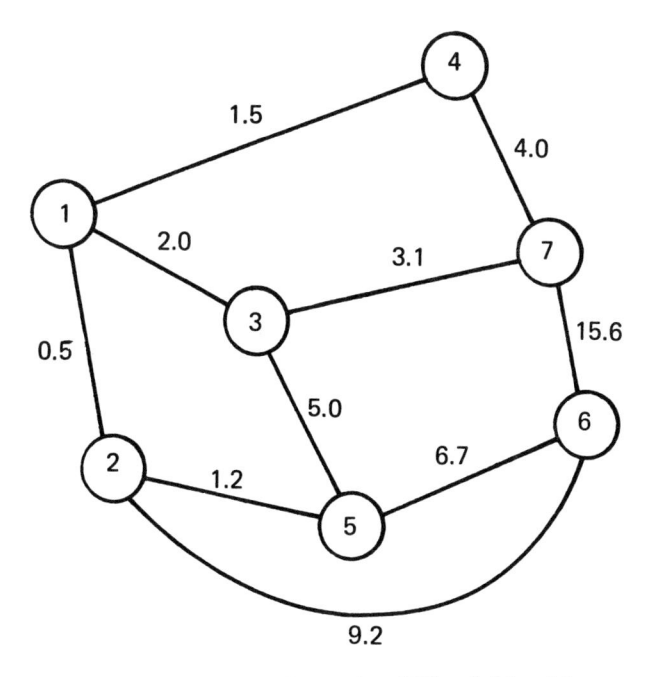

Figure 3–2. Example of Floyd Algorithm

Remarks on Applications

If only one link ends at a node, as is the case for node x in figure 3-3, it is evident that the length of the shortest route from x to any other node z is $d_{xy} + s_{yz}$, and that $r_{xz} = y$. Thus nodes such as x, called *pendant* nodes, may be eliminated in performing the algorithm, and the necessary S and R elements for them calculated afterward.

Somewhat similar reasoning may be applied to eliminate nodes at which only two links end. Suppose that only links (j, i) and (i, k) end at node i. If $d' = d_{ji} + d_{ik} > d_{jk}$, then node i may be removed without affecting the results of the algorithm for nodes other than i. If $d' < d_{jk}$, then node i may be removed with a similar result, after replacing d_{jk} by d'; of course, any occurrence of link (j, k) must thereafter be understood as meaning route $(j, i), (i, k)$. Once the algorithm has been performed for the modified graph, viz., the original graph with node i removed (see section 1-2), the length of the shortest route from i to any other node z is $s_{zj} + d_{ji}$ or $s_{zk} + d_{ik}$, according to whether $d_{ji} < d_{ik}$ or the reverse is true. Whereas this description is in terms of the Floyd algorithm, the technique of node elimination and length substitution has been extended to create an alternative algorithm [7] which, for sparse networks (relatively few links), is more efficient than Floyd.

Hu gives an efficient method, that of *decomposition*, of using the Floyd algorithm to find all shortest-route lengths in sparse networks [8]. The set of

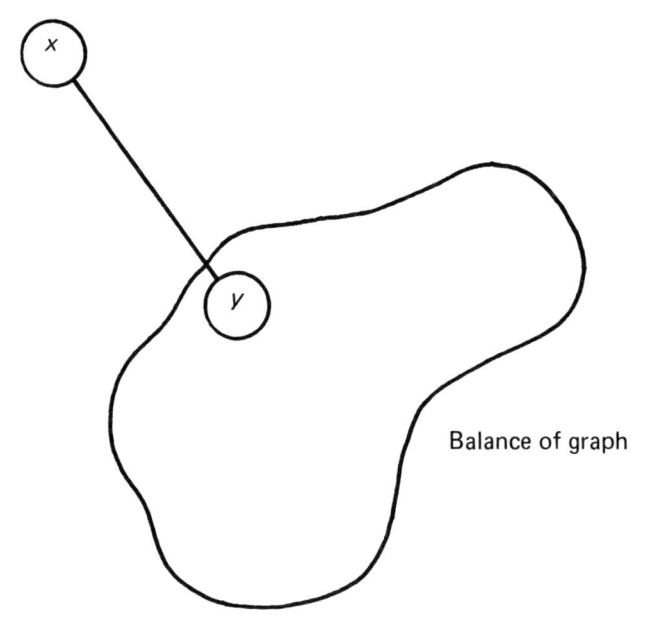

Figure 3-3. Graph in which Only One Link Ends at Node x

nodes, N, is partitioned into three disjoint subsets A, X, and B such that $A \cup X \cup B = N$. These subsets are defined as follows. Consider A for the moment as an arbitrary subset of N. Then X is the set of *neighbors* of nodes in A, that is, the set of nodes[a] $\{i: i \notin A$ and $d_{ij} < \infty$ for $j \in A\}$. Finally, $B = N - (A \cup X)$. In practice, A and B are chosen as relatively large subsets of N, with X relatively small. Hu applies the Floyd algorithm separately to the subgraphs of G whose node sets are A, B, X, $A \cup X$, and $B \cup X$, and demonstrates how the results may be used to find S, the matrix of shortest-route lengths for all node pairs in G. In this way, the total number of computations that are the core of the Floyd algorithm (see step 5) can be substantially reduced, compared to the number when the algorithm is applied directly to G, provided that G is sparsely connected and that A and B are judiciously chosen.

3-4 Shortest Routes from a Specified Node to All Other Nodes

The lengths of the shortest routes from a specified node to all other nodes are included in the results of the Floyd algorithm. Nevertheless, it is interesting to consider the specified-node problem separately since it introduces the class of *labeling* algorithms, which are useful in some situations.

Dijkstra Algorithm (Lengths of Shortest Routes from a Specified Node to All Other Nodes [6, 9])

The specified node is here taken as node 1 for convenience. For graph $G(N, L)$, find $s_{1j}, j = 2, \ldots, n$, the lengths of the shortest routes from node 1 to nodes $2, \ldots, n$.

Step 1 [Initialization]. By the expression "label node i with x," we mean "set node weight $w_i \leftarrow x$." Label node 1 permanently with zero. Label nodes $2, \ldots, n$ tentatively with ∞. Let N_P be the subset of permanently labeled nodes of N; initially, $N_P = \{1\}$. (A permanent label is one whose value, once set, remains the same throughout the balance of the algorithm. A tentative label is one whose value may be changed.)

Step 2 [Update the labels of nodes in $N - N_P$]. For $i \in N - N_P$ and $j \in N_P$, label i tentatively with min $(w_i, w_j + d_{ji})$.

---

[a] It can be shown that X is a node cut-set of G (see section 1-2).

Step 3 [Permanently label one or more nodes in $N - N_P$]. Let $w^* = \min(w_i: i \in N - N_P)$. Make the label of any node tentatively labeled with w^* permanent. If all nodes are now permanently labeled, stop; the label of node j is now $s_{1j}, j = 2, \ldots, n$. Otherwise, set $N_P \leftarrow N_P \cup$ {nodes permanently labeled at this step} and go to step 2.

As an example, we will apply the algorithm to figure 3–2. In the first pass through step 2, we have $w_2 = 0.5$, $w_3 = 2.0$, and $w_4 = 1.5$; the values of w_5, w_6, and w_7 remain at ∞. Thus at step 3, the label of node 2 is made permanent. In the next pass through step 2, we find revised values $w_5 = 1.7$ and $w_6 = 9.7$; the minimum value among the tentatively labeled nodes is $w_4 = 1.5$, so the label of node 4 is made permanent. The complete algorithm requires $n - 1 = 6$ passes through steps 2 and 3 altogether, and it is easily seen that nodes $2, \ldots, 7$ at the end are labeled with $s_{12}, \ldots, s_{17}$ as found previously in section 3–3.

To find the shortest routes between node 1 and the other nodes, it is necessary only in performing the algorithm to maintain a table that shows for each node $m \geq 2$, with permanent label w_m, its predecessor node, that is, the (permanently labeled) node j for which $w_j + d_{ji} = w_m$ at step 2. In the example of figure 3–3, the table is as follows:

Node	Predecessor Node
2	1
3	1
4	1
5	2
6	5
7	3

To find the shortest route from node 1 to node 7, for example, we proceed backwards from 7 to its predecessor 3, and then from 3 to its predecessor, 1; the route is (1, 3), (3, 7). A node may have more than one predecessor node, in which case there are additional tables, each corresponding to a shortest route. Thus the algorithm identifies all shortest routes, in the event of ties.

To find the lengths of the shortest routes from node 1 to all nodes in a subset, N', of $N - \{1\}$, we simply stop the algorithm at step 3, when all nodes in N' are permanently labeled. In figure 3–3, for example, only one pass through steps 2 and 3 is necessary to determine that the shortest route from node 1 to node 2 has length 0.5. Enough predecessor node information will exist when the algorithm is truncated in this way to determine the routes from 1 to all nodes in N'.

Routes   29

By specifying nodes 1, 2, ..., n in succession, and performing the Dijkstra algorithm for each node, we can find the lengths of the shortest routes for all node pairs, as well as the relevant route-identifying tables. As Dreyfus shows, this takes more computation time than Floyd, when the all-pairs problem is at issue [6].

Suppose that for each $j \neq i$, P_j is a shortest route from i to j. We can associate with P_j a subgraph G_j in the following obvious way: G_j consists of the links in P_j, (i, i_1), ..., (i_k, j), and the nodes $i, i_1, ..., i_k, j$. Then for any i, the union of the G_j, that is,

$$T_i = \bigcup_{j \in N, j \neq i} G_j$$

is a *shortest-route tree* for i. [A shortest-route tree is not in general the same as a minimal spanning tree, or tree of minimum weight (see chapter 2).] That is, T_i is a tree on N such that the unique route in T_i from i to j is a shortest route. For a specified i, finding a shortest-route tree evidently solves the shortest-route problem, both as to lengths and the routes themselves. Gilsinn and Witzgall have studied labeling algorithms for finding shortest-route trees, again with large, sparsely connected networks in mind [10]. They give flowcharts and FORTRAN programs for seven algorithms, as well as comparative computation-time statistics for these algorithms applied to networks with 100 to 3,000 nodes. In addition, they recommend methods of economizing on storage, relative to matrix methods, such as the Floyd algorithm.

For large networks, particularly if they are sparsely connected, the designer may wish to consider labeling methods or may need to consider them if storage is the main limitation. Their greater efficiency in this regard may offset the complications of programming they require, as compared to matrix methods, for example, the Floyd algorithm.

3-5 Generating Longer Routes

It is often desirable to be able to generate routes longer than the shortest route, for example, as alternates to the latter in switched networks when the shortest route is unavailable for some reason. It is logical to approach the generation of longer routes by asking for the kth shortest routes, for $k = 2, 3, ...$, that is, the second-, third-, ..., shortest routes. This problem has received some attention, though not so much as the shortest-route problem ($k = 1$), both for the specified-node-pair case [6] and the all-node-pairs case [11].

In many practical problems, what is desired rather than to identify the kth shortest routes is to find all routes between nodes i and j whose length does not exceed M, where $M \geq s_{ij}$, the length of the shortest route. If $M = s_{ij}$, then only the shortest route(s) will do, so in general we will have $M > s_{ij}$; we can assume that M is large enough so that there are routes of interest. For example, if $d_{rs} = 1$ for every feasible link (r, s), then we ask for all routes that contain M or fewer links. As another example, if $M = s_{ij} + b$, with $b > 0$, we ask for all routes whose length does not exceed that of the shortest by more than b; instead of a fixed increment, we could have b_{ij}.

For convenience, we will call a route of length $\leq M$ *acceptable*. The author has used a procedure for finding all acceptable routes between i and j based on the following principle. Suppose that a route P_k of length C_k can be found from i to k and that $C_k + s_{kj} \leq M$. Then there may be an acceptable route from i to j that contains P_k as a subset. We cannot say for sure that there is such a route since concatenating P_k with a suitable route from k to j may produce a set of links that contains a cycle, that is, that is not a route at all. Of course, there may be more than one acceptable route that contains P_k. On the other hand, if $C_k + s_{kj} > M$, there is no acceptable route of this kind. This principle is embodied in the following algorithm.

Route-tracing Algorithm (Acceptable Routes)

For graph $G(N, L)$, find all acceptable routes between node i and node j and their lengths.

Step 1 [Initialization]. The algorithm requires the shortest-route-length matrix S, which can be found using the Floyd algorithm, section 3-3. Let m be a node index, and set $m \leftarrow 1$ and $i_m \leftarrow i$. Let P be a partial or test route, and set $P \leftarrow \phi$, the empty set. Let C be a cumulative route length, and set $C \leftarrow 0$.

Step 2 [Search for next link]. Let i_a be a node directly connected to i_m such that $P \cup \{(i, i_a)\}$ does not contain a cycle and $C + d_{i_m i_a} + s_{i_a j} \leq M$. Set $P \leftarrow P \cup \{(i_m, i_a)\}$ and go to step 4. (Node x is said to be directly connected to node y if there is a link between x and y, that is, if $d_{xy} < \infty$.) If all nodes directly connected to i_m have been tried, go to step 3.

Step 3 [End algorithm or backtrack]. If $m = 1$, stop. Otherwise, set $m \leftarrow m - 1$ and go to step 2.

Step 4 [Test for complete route or continue] If $i_a = j$, save P and C, set $P \leftarrow \phi$ and $C \leftarrow 0$, and go to step 2. If $i_a \neq j$, set $C \leftarrow C + d_{i_m i_a}$, $m \leftarrow m + 1$, $i_m \leftarrow i_a$, and go to step 2. (The expression "save P and C" means, in computer terms, that we write the list of links making up P and the value of C on an output medium, or store them in memory, for use as part of the results of the algorithm.)

Suppose, for example, that we have $M = 7.0$ in figure 3-2, $i = 1$, and $j = 5$. If, at step 2, we try $i_a = 2$, that is, link (1, 2) as the first in P, we have $C + d_{12} + s_{25} = 0 + 0.5 + 1.2 = 1.7 \leq M$, using the S matrix in section 3-3. Thus, (1, 2) is the first link of the trial route. At step 4, we obtain $C = 0.5$, $m = 2$, and $i_2 = 2$. As we continue, we find that, of the two nodes directly connected to 2, other than 1, one results in the acceptable (and shortest) route $\{(1, 2), (2, 5)\}$ and the other fails since $d_{12} + d_{26} + s_{65} = 0.5 + 9.2 + 6.7 > M$.

Continuing this example, if we try $i_a = 3$ as the next node directly connected to 1, we have, at step 2 with $m = 1$ once more, and $C = 0$, $C + d_{13} + s_{35} = 0 + 2.0 + 3.7 \leq M$, and link (1, 3) is the first in P. Here it is not the shortest route from 3 to 5, viz., $\{(3, 1), (1, 2), (2, 5)\}$, that is of use but rather route $\{(3, 5)\}$, and the acceptable route via 3 is $\{(1, 3), (3, 5)\}$. When node 4 is considered as the final node directly connected to 1, we have $C + d_{14} + s_{45} = 0 + 1.5 + 3.2 \leq M$, but there is actually no acceptable route via 4.

The route-tracing algorithm is simple to program and, notwithstanding the false starts that occur, as illustrated in the example, it is reasonably efficient. The S matrix is generally needed in estimating a value, or set of values, for M, so there is no loss in calculating it for the initialization. The routes are not generated in any systematic order as to length, so that a sort is needed, after the algorithm is performed, to get a list of routes in order of increasing length or other required order.

3-6 Node-disjoint and Link-disjoint Routes

If each route in some set of routes between nodes i and j passes through one or more of the same intermediate nodes, service on all the routes is interrupted if messages cannot get through any of the common nodes, by reason of damage or unavailability. Similarly, if several routes contain one or more links in common, service on all the routes is interrupted if any of the common links fails. We are thus led to seek *node-disjoint* routes, that is, those that pass through no common nodes except i and j and *link-disjoint* routes, that is, those that contain no links in common. Two routes that are node-disjoint are link-disjoint, but the converse is not necessarily true.

We generally require not only that the routes have a disjointness property but also that they meet some additional constraints. For example, if we need just two routes between nodes i and j, we may wish to use (1) the shortest *(a)* Original graph. Shortest route between 1 and 5 is $\{(1, 3), (3, 5)\}$.

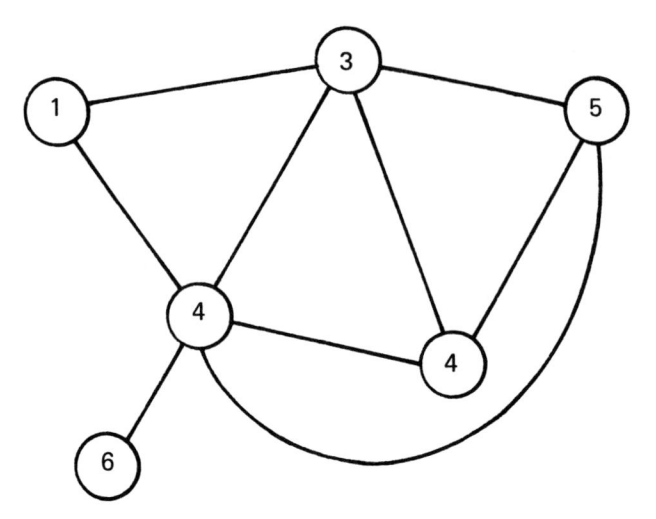

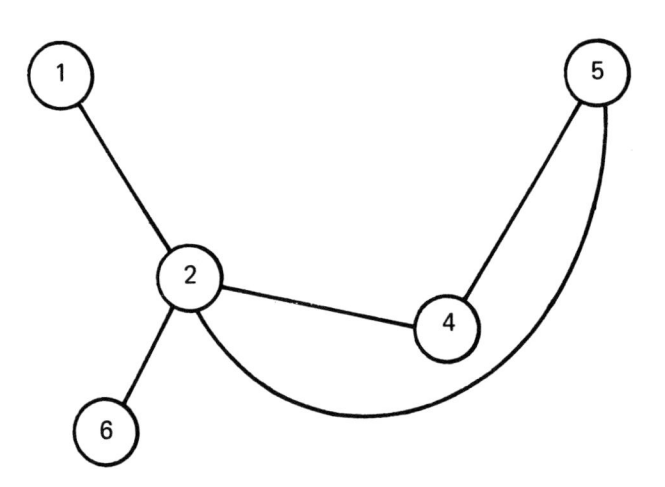

(b) Reduced graph.

Figure 3-4. Example of Node-Disjoint Routes

route, say P_s, and (2) the shortest route that is node-disjoint from P_s, say P_d. Route P_s can be found by applying one of the algorithms in sections 3-3 or 3-4. Klemushin has shown that P_d may be found by deleting from $G(N, L)$, the original graph, all intermediate nodes in P_s, that is, all nodes traversed by P_s (i and j excluded), resulting in a reduced graph $G'(N', L')$, and then finding the shortest route between i and j in G' [12]. For example, if in figure 3-4(a) the shortest route between nodes 1 and 5 is $P_s = \{(1, 3), (3, 5)\}$, then in the reduced graph shown in figure 3-4(b), P_d equals the shorter of $\{(1, 2), (2, 4), (4, 5)\}$ or $\{(1, 2), (2, 5)\}$. This process can be carried further to find route P'_d, the shortest route that is node-disjoint from both P_s and P_d, and so on, if the network permits. Figure 3-4(b), for example, cannot be reduced further for routes between nodes 1 and 5; no reduction of the original network, figure 3-4(a), is possible if pendant node 6 is one of the route ends. Klemushin describes efficient procedures for the all-node-pairs and specified-node-pair cases of this extended problem.

Suurballe addresses the problem of finding two or more node-disjoint routes between a specified pair of nodes whose *total* length is minimum [13], a constraint that does not necessarily lead to the same results as that of Klemushin. The enumeration and generation of routes with disjointedness properties are considered further in chapter 4.

References

1. Rawdin, E., and Bedrosian, S. D. "On Enumerating Paths of K Arcs in Unoriented Complete Graphs," *J. Franklin Inst.* 299:73-76, 1975.
2. Pierce, A.R."Bibliography on Algorithms for Shortest Path, Shortest Spanning Tree, and Related Circuit Routing Problems," *Networks* 5:129-149, 1975.
3. Shek, C.H. "Constrained Shortest Route Paths in Networks." Dissertation, Montana State University, 1975. UM NO. 76-4875.
4. Floyd, R.W. "Algorithm 97: Shortest Path," *Commun. ACM* 5:345, 1962.
5. Frank, H., and Frisch, I.T. "The Design of Large-scale Networks," *Proc. IEEE* 60:6-11, 1972.
6. Dreyfus, S.E."An Appraisal of Some Shortest-path Algorithms," *Operations Res.* 17:395-412, 1969.
7. Network Analysis Corporation. Final Technical Report... for the Project, "Research in Store and Forward Computer Networks." AD-757090, 1972.
8. Hu, T.C. "A Decomposition Algorithm for Shortest Paths in a Network," *Operations Res.* 16:91-102, 1968.

9. Dijkstra, E.W. "A Note on Two Problems in Connexion with Graphs," *Numerische Math.* 1:269–271, 1959.
10. Gilsinn, J., and Witzgall, C. "A Performance Comparison of Labelling Algorithms for Calculating Shortest Path Trees," *NBS Technical Note 772*, 1973.
11. Minieka, E. "On Computing Sets of Shortest Paths in a Graph," *Commun. ACM* 17:351–353, 1974.
12. Klemushin, G.N. "Computation of Node-disjoint Paths for Communication Network Planning." Dissertation, University of Pennsylvania, 1971. UM No. 72-17378.
13. Suurballe, J.W. "Disjoint Paths in a Network," *Networks* 4:125–145, 1974.

4 Reliability of Networks

4-1 Introduction

A variety of measures of network reliability, that is, of the ability of a network to continue to afford communications routes between some nodes when other nodes or links fail, have been proposed. These measures fall into two classes: *Deterministic* measures depend only on the structure of the network, that is, on the numbers of nodes and links and the way they are connected. *Probabilistic* measures of availability, on the other hand, depend not only on the structure but also on the probabilities of failure of nodes and links. In a major review paper, Wilkov describes several measures in both classes [1].

In this chapter we consider analytic methods for finding two of the basic deterministic measures and also an important probabilistic measure, the terminal reliability. Many valuable additional results in network reliability have been found by simulation, which is beyond the scope of this book.

4-2 Cohesion and Connectivity

If i and j are distinct nodes of a connected graph $G(N, L)$, then an i, j *link cut-set* of G is one whose removal results in a graph with two components, one containing i and the other containing j. Suppose that v_{ij} is the minimum number of links in any i, j link cut-set; thus v_{ij} is the minimum number of links that must be removed from G to break all routes between i and j. Furthermore, v_{ij} is the maximum number of link-disjoint routes between i and j [1]. The *cohesion* of G is the minimum of v_{ij} for all node pairs and is denoted by v. It is the minimum number of links that must be removed from G to break all routes between at least one pair of nodes and is accordingly one of two basic measures of network reliability.

The other measure is the *connectivity*, which is similarly defined in relation to node cut-sets of G. An i, j *node cut-set* of G is one whose removal results in a graph in which i and j are in separate components. If ω_{ij} is the minimum number of nodes in an i, j cut-set and thus is the minimum number of nodes that must be removed from G to break all routes between i and j (it is also the maximum number of node-disjoint routes between i and j [1]), then the connectivity of G, denoted by ω, is the minimum of ω_{ij} for all node pairs. [Only node cut-sets that contain neither i nor j are considered in finding ω_{ij}.

35

If G is a *complete* graph, that is, one in which all links (i,j) are present, ω_{ij} is undefined.] It is the minimum number of nodes that must be removed from G to break all routes between at least one pair of nodes.

The quantities of ω_{ij} and v_{ij} are related by the inequality $\omega_{ij} \leq v_{ij}$; thus $\omega \leq v$, and it is never necessary to remove more nodes than links to disconnect G. Further, v cannot exceed d^*, the minimum degree of any node of G. Since d^* is in turn bounded by the average degree $[2l/n]$, where l and n are the numbers of links and nodes in G and $[x]$ means the largest integer $\leq x$, we have $\omega \leq v \leq d^* \leq [2l/n]$ [1].

4–3 Finding the Cohesion of an Undirected Graph

The basic algorithm for finding the cohesion v is an application of a famous theorem, the max-flow min-cut theorem [2]. To understand this algorithm, we must first define a *directed graph* and certain concepts related to it since the algorithm is stated for this kind of graph. Then we show how the algorithm is used to find the cohesion of an undirected graph.

Directed Graphs and the Max-flow Min-cut Theorem

A directed graph is one in which, contrary to our general rules (section 1–1), each link (i,j) has a direction, that is, an order of the pair i,j; link (i,j) is thus not the same as link (j,i), and in fact, both may exist in the same graph. Links (i,i) are excluded, as in the case of undirected graphs. Figure 4–1 shows a directed graph, with the direction of each link indicated by an

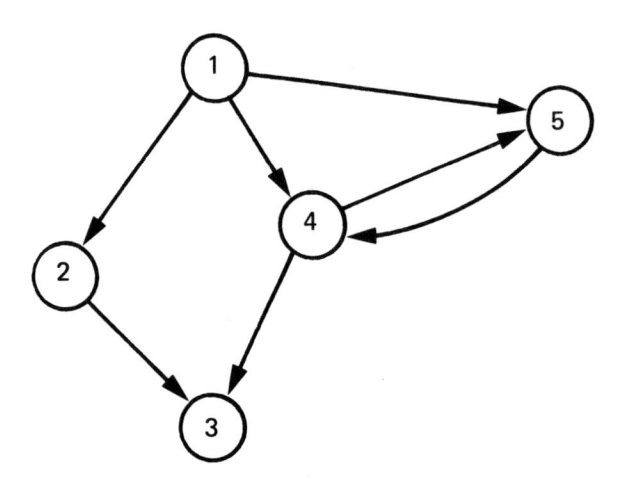

Figure 4–1. Directed Graph

Reliability of Networks

arrowhead; link (1, 4), for example, is directed from node 1 to node 4. We will denote a directed graph with the set of nodes N and the set of links L^* by $G^*(N, L^*)$. A *directed route* in G^* is a defined similarly to a route in an undirected graph, except that it must follow the link directions; in figure 4–1, a directed route between node 1 and node 3 is $\{(1, 5), (5, 4), (4, 3)\}$.

Suppose that A is a subset of N and A' is its complement, that is $A' = N - A$. A *cut* (A, A') in G^* is the set of links of G^* that are directed from a node in A to a node in A^*; for example, in figure 4–1, if $A = \{1, 2\}$, then cut $(A, A') = \{(1, 4), (1, 5), (2, 3)\}$. If node i is in A and node j is in A', the cut (A, A') is called an i, j cut. The removal from G^* of the links in an i, j cut breaks every directed route between i and j. Thus the cut (A, A') previously described for figure 4–1 is a 1, 3 cut, and its removal breaks every directed route from 1 to 3; it is also a 1, 4 cut, and so forth.

A *directed i, j link cut-set* in G^* is a set of links that has the same property as an i, j cut, that is, its removal from G^* breaks every directed route between i and j, but it is, in addition, minimal. When it is clear that we are dealing with a directed graph, we will refer simply to an i, j link cut-set, dropping "directed." The cut (A, A') described in the previous paragraph for figure 4–1 is a 2, 3 cut, but it is not a 2, 3 link cut-set since the removal of the smaller set of links $\{(2, 3)\}$ breaks every directed route between node 2 and node 3; the latter is, in fact, a 2, 3 link cut-set.

A *link flow* from i to j in G^*, f_{ij}, is a nonnegative link weight; it is not necessary that $f_{ij} = f_{ji}$. We will assume $f_{ii} = 0$ for directed graphs, as in the case of undirected graphs. Similarly, the *capacity* of link (i, j), c_{ij}, is another nonnegative link weight; it is not necessary that $c_{ij} = c_{ji}$. A *feasible s, t flow pattern* in G^* is a set of link flows that satisfies the following equations for distinct nodes s and t:

$$\sum_{x \in N} f_{ix} - \sum_{y \in N} f_{yi} = \begin{cases} v_{st} & \text{if } i = s \\ 0 & \text{if } i \neq s \text{ or } t \\ -v_{st} & \text{if } i = t \end{cases} \quad (4.1)$$

$$c_{ij} \geq f_{ij} \geq 0 \; i, j \in N \quad (4.2)$$

The first term on the left in equation 4.1 is the total flow out of i, and the second term is the total flow into i. Thus equation 4.1 expresses the "conservation of flow," with s being the "source" and t the "sink." The nonnegative quantity v_{st} is the *value* of the flow pattern. Figure 4–2 shows a feasible flow pattern for $s = 1$ and $t = 4$, with the value $v_{14} = 3$.

The max-flow min-cut theorem says that, for any distinct nodes s and t in directed graph G^*, the maximum value of a feasible s, t flow pattern is equal to the minimum capacity of an s, t cut. [2] (The capacity of any set of links

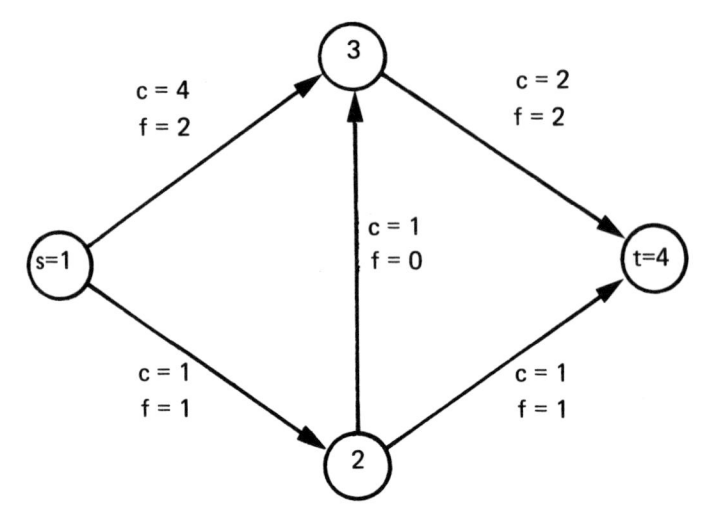

Figure 4-2. Directed Graph with Feasible Flow Pattern

such as an s, t cut is the sum of the capacities of the links in the set.) In figure 4–2, for example, the 1, 4 cuts $\{(1, 2), (3, 4)\}$ and $\{(2, 4), (3, 4)\}$ each have capacity 3, and it is easy to show that this is the minimum capacity of any 1, 4 cut. Thus the value of the feasible 1, 4 flow pattern shown is the greatest that can be achieved.

Ford and Fulkerson give an algorithm for finding the maximum s, t flow in a directed graph G^*, that is, the maximum value V_{st} of a feasible s, t flow pattern, and hence the minimum capacity of an s, t cut, given that the link capacities are integers [2]. The following algorithm is a special case of that of Ford and Fulkerson, in which each link has capacity 1, and each link flow may be either 0 or 1; it follows that the capacity of a cut is the number of links it contains. Because the minimum s, t cut capacity is equal to the minimum s, t link cut-set capacity, we thus have V_{st} equals the number of links in a minimum s, t link cut-set of G^* [3].

Labeling and Augmentation Algorithm
(Number of Links in a Minimum s, t Link Cut-set) [3, 4]

For directed graph $G^*(N, L^*)$, find V_{st}, the number of links in a minimum s, t link cut-set.

Step 1 [Initialization] Each link in L^* is understood to have capacity 1. For each link in L^*, define a flow f_{ij} which may have the value 0 or 1. Initially, set

$f_{ij} \leftarrow 0$ for all (i, j) in L^*. A node is said to be labeled if it has been assigned a weight, that is, a pair of values $(x, \pm)$ where x is either a node number or zero; the meaning of the second part of the label, $+$ or $-$, is given later. Initially, all nodes are unlabeled. Also, all nodes are initially unscanned; the meaning of a node's having been scanned is given later.

Step 2 [Begin labeling]. Label s with $(0, +)$.

Step 3 [Scan a labeled node]. Select any labeled and unscanned node i. For all unlabeled nodes y such that link (y, i) has flow $f_{yi} = 1$, label y with $(i, -)$; for all unlabeled nodes y such that link (i, y) has flow $f_{iy} = 0$, label y with $(i, +)$. When this step has been completed, node i is said to be scanned.

Step 4 [Continue scanning]. Repeat step 3 until t is labeled and unscanned or until no more labels can be assigned and t is unlabeled. In the latter case, stop;

$$V_{st} = \sum_{y:(s,y) \in L^*} f_{sy}$$

In the former case, go to step 5.

Step 5 [Initialize for flow change]. Let z be a node number and set $z \leftarrow t$. Let q and r be node numbers whose values are at the moment unassigned.

Step 6 [Change flow]. If z is labeled with $(q, +)$, set $f_{qz} \leftarrow f_{qz} + 1$. If z is labeled with $(q, -)$, set $f_{qz} \leftarrow f_{qz} - 1$. Set $r \leftarrow q$.

Step 7 [Continue to change flow unless flow pattern is complete, in which case return to labeling]. If $r = s$, erase all labels and return to step 2. Otherwise, set $z \leftarrow r$ and go to step 6.

Figure 4–3 illustrates the algorithm. At the first pass through step 3, only node s is labeled and unscanned, and as a result, both x and y are unlabeled $(s, +)$. Continuing, node t is labeled $(x, +)$; after steps 5, 6, and 7, the link flows are as shown in figure 4–3(b), and after another pass through steps 2 through 7, they are as shown in figure 4–3(c). When we next reach step 3, we find that no more labels can be assigned since s is the only labeled node and $f_{sx} = f_{sy} = 1$; thus the algorithm terminates with $V_{st} = 2$. Figure 4–3(c) shows that the algorithm not only gives the value of V_{st} but also identifies the two link-disjoint routes from s to t, viz. $\{(s, x), (x, t)\}$ and $\{(s, y), (y, t)\}$.

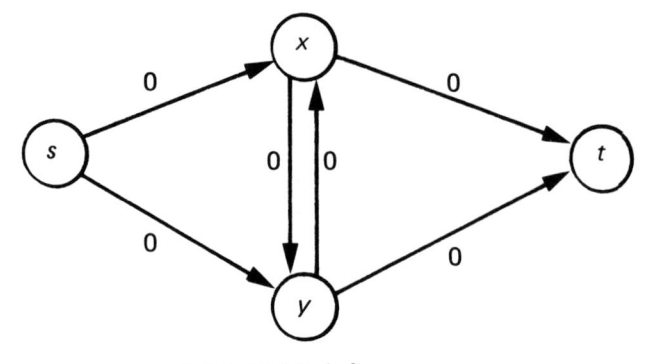

(a) Initial link flows.

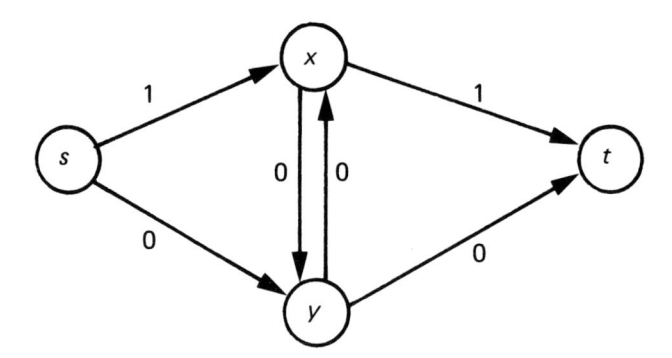

(b) Link flows after first pass through Step 2.

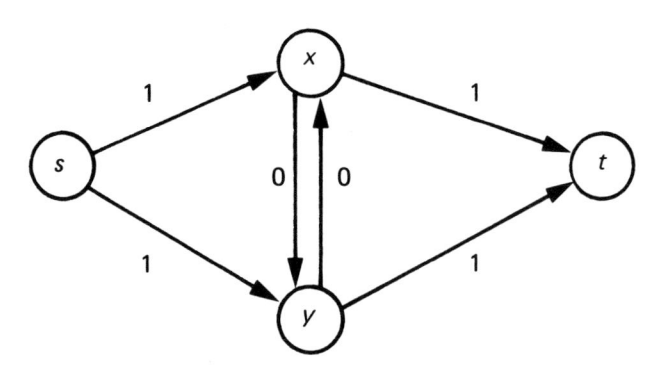

(c) Link flows after second pass through Step 2.

Figure 4-3. Example of Labeling and Augmentation Algorithm

Reliability of Networks 41

Application to Undirected Graphs

Starting with undirected graph $G(N, L)$, we choose a particular pair s, t for which we wish to find v_{st}, and proceed as follows. First, we derive from $G(N, L)$ a directed graph $G^*(N, L^*)$ by (a) replacing each link that ends at s in L by a directed link in L^* from s to the appropriate node; (b) replacing each link that ends at t in L by a directed link in L^* from the appropriate node to t; and (c) replacing each other link (i, j) in L by two directed links (i, j) and (j, i) in L^*. This transformation is illustrated in figure 4–4 [5]. Now we apply the labeling and augmentation algorithm to G^* and find V_{st} equals the number of links in a minimum s, t link cut-set of G^*. This is also the number of links in a minimum s, t link cut-set of G and is therefore the value of v_{st} in G.

Owing to symmetry, $v_{ij} = v_{ji}$ in G. Thus we need apply the labeling and augmentation algorithm to at most $n(n - 1)/2$ node pairs to find v, the cohesion of G; for example, if we choose those pairs i, j with $i < j$, we have $v = \min(v_{ij}: i < j)$. For each node pair of G to which the algorithm is applied, G is transformed into a different directed graph, following rules a, b, and c of the preceding paragraph. In fact, there are only $n - 1$ independent values of v_{st} for G, and an algorithm of Gomory and Hu can be applied to find v with only $n - 1$ iterations of the labeling and augmentation algorithm [4, 6].

4–4 Finding the Connectivity of an Undirected Graph

To find ω_{st}, the node-based (*connectivity*) problem can be converted into a link-based (*cohesion*) problem, as follows. First, we derive from $G(N, L)$, the original undirected graph, a directed graph $G^*(N, L^*)$, as in section 4–3. Next we derive from $G^*(N, L^*)$ in turn a new directed graph $\hat{G}(\hat{N}, \hat{L})$ by (a) splitting each node i of N, except s and t, into two new nodes i_1 and i_2 in $\hat{N}$, and connecting these with the directed link (i_1, i_2) in $\hat{L}$; (b) replacing each link (j, i) in L^* by the link (j_2, i_1) in $\hat{L}$. In carrying out rule b, we imagine that s is renumbered s_2 and t is renumbered t_1 in $\hat{N}$. Figure 4–5 shows the result of transforming graph G^* of figure 4–4(b) in this way. In $\hat{G}$, a route from $s = s_2$ to $t = t_1$ must now include links between the parts of a split node such as link (x_1, x_2); the inclusion of this link in a link cut-set has the same effect on breaking any directed route from s_2 to t_1 in $\hat{G}$ as does the removal of node x in the original undirected graph G. It can be shown that the cohesion of $\hat{G}$, $v(\hat{G})$, is equal to the connectivity of G, $\omega(G)$. Since $v(\hat{G})$ can be found with the labeling and augmentation algorithm, the problem is solved, albeit with greater complexity of computation than in finding $v(G)$, owing to the increased numbers of nodes and links in $\hat{G}$ as compared to G^*. The labeling and augmentation algorithm, when applied to $\hat{G}$, again not only gives the value of $v(\hat{G})$ but also identifies the equivalent number of node-disjoint routes from s to t.

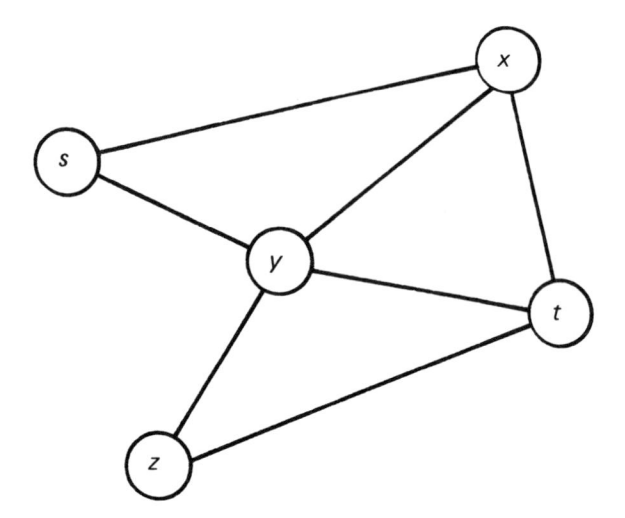

(a) Original undirected graph G(N,L).

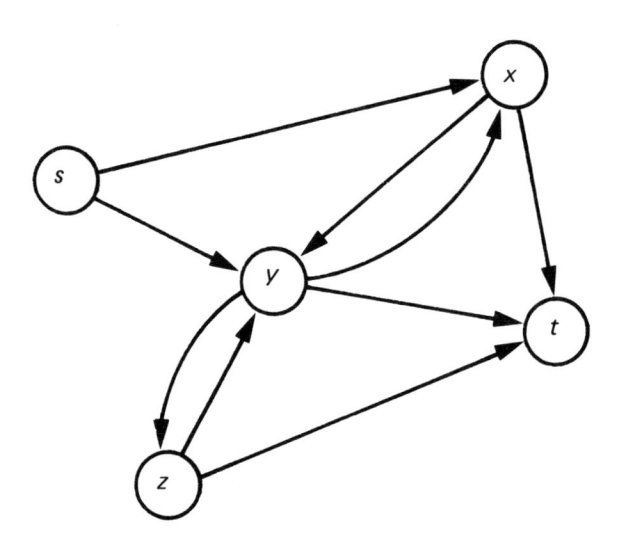

(b) Directed graph G*(N,L*).

Figure 4–4. (*a*) Original Undirected Graph $G(N, L)$ (*b*) Directed Graph $G^*(N, L^*)$

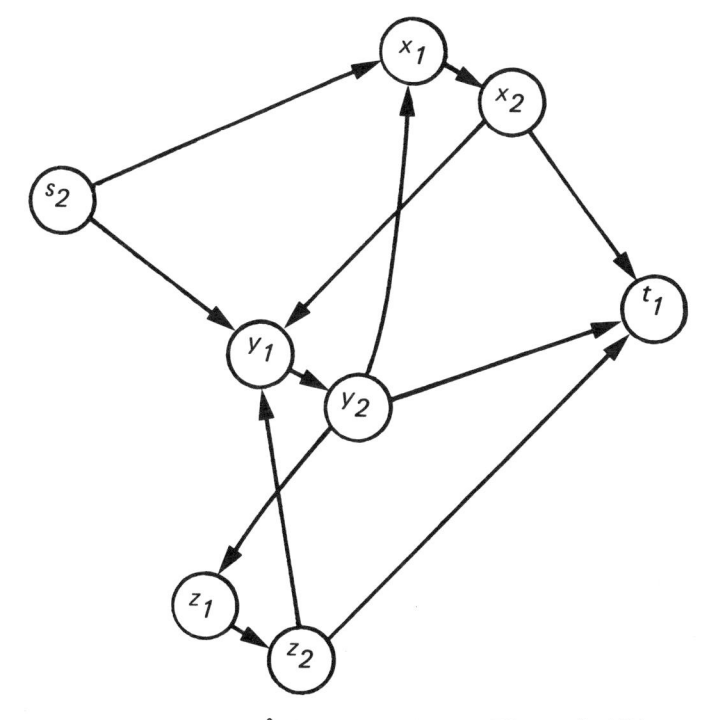

Figure 4-5. Graph $\hat{G}$ Corresponding to Figure 4-4(b)

An alternative method for finding ω, the $\omega_{s,t}$ algorithm of Frank [3], is an adaptation of the labeling and augmentation algorithm. Since it avoids the need to split nodes, as in graph $\hat{G}$, and deals directly with graph G^*, it is more efficient than the previous method.

m-Connectivity

A more limited problem is to show that for the undirected graph G, ω is bounded below, that is, $\omega \geq m$; G is then said to be *m*-connected [1]. To begin with, an upper bound on ω is given by the minimum node degree, d^* (section 4-2). Thus if the degree of each node is not at least m, G cannot be *m*-connected.

To go further, we may use a theorem proven by Kleitman, which allows us to deal with successively smaller networks and is as follows [7]. Graph G is *m*-connected if the following steps can be completed for graphs $G, G_1, G_2,$ $\ldots, G_{m-1}$: (1) Select any one node, say i_0, in G, and show that there are m node-disjoint routes from i_0 to all other nodes of G; (2) remove i_0 from G [we recall that removing a node from a graph means deleting the node itself as well as all links that end at the node (section 1–2).], forming graph G_1, and show that, in G_1, there are $m - 1$ node-disjoint routes from any one node, say i_1, to all other nodes; (3) form graph G_2 by removing i_1 from G_1, and show that in G_2 there are $m - 2$ node-disjoint routes from any one node i_2 to all other nodes, and so on. If G is m-connected, we will be able to find the necessary number of node-disjoint routes for each successive graph $G, G_1, \ldots, G_{m-1}$; the last step, for G_{m-1}, will be simply to show that there is one route in G_{m-1} from some node to all other nodes. If, on the other hand, G is not m-connected, that is, if $\omega < m$, the iterative process will fail for one of the graphs $G, G_1, \ldots, G_{m-1}$; it need not be carried beyond the step at which it fails.

An example is shown in figure 4–6 [8]. To show that the original network, figure 4–6(a) is 4-connected, select node 1, which has at least four node-disjoint routes to each other node. Remove node 1, resulting in the network of figure 4–6(b). Continuing the Kleitman procedure, we next remove node 2, resulting in figure 4–6(c), and finally node 3, resulting in figure 4–6(d). Since the latter network is connected, there is at least one route from any node to any other node, and we have established that the network of figure 4–6(a) is 4-connected.

We may perform each step required by Kleitman's theorem by finding, for graph $G_j(G_0 = G)$, the $n - j - 1$ connectivities $\omega_{ijk}, k \in \{N - i_0 - i_1 - \cdots - i_j\}$, where n is the number of nodes in N, using the methods described earlier in this section. If we denote by ω_j the minimum of these values for graph G_j, we must have $\omega_j \geq m - j$. This will require finding the connectivities for at most $nm - m(m + 1)/2$ node pairs altogether, as compared to $n(n - 1)/2$ node pairs if the entire graph of G were to be tested. For example, we can establish that a 10-node graph is 3-connected by checking 24 node pairs instead of 45. Alternatively, additional theorems of Kleitman's can be applied to reduce the amount of computation at each step [7].

4–5 Probabilistic Link and Node Failures

The links and nodes of a network may fail to operate because of equipment breakdowns, natural disasters, or hostile action. We will assume that each link (i, j) has a probability of failure p_{ij}; that each node i has a probability of failure p_i; and that all the link and node failures are statistically independent. We address the problem of calculating the *terminal reliability* for a particular node pair s, t that is, the probability Q_{st} that at least one operable route exists from s to t.

We illustrate one approach to finding Q_{st} with the network shown in figure 4–7. To simplify the explanation, assume that all the node-failure prob-

Reliability of Networks

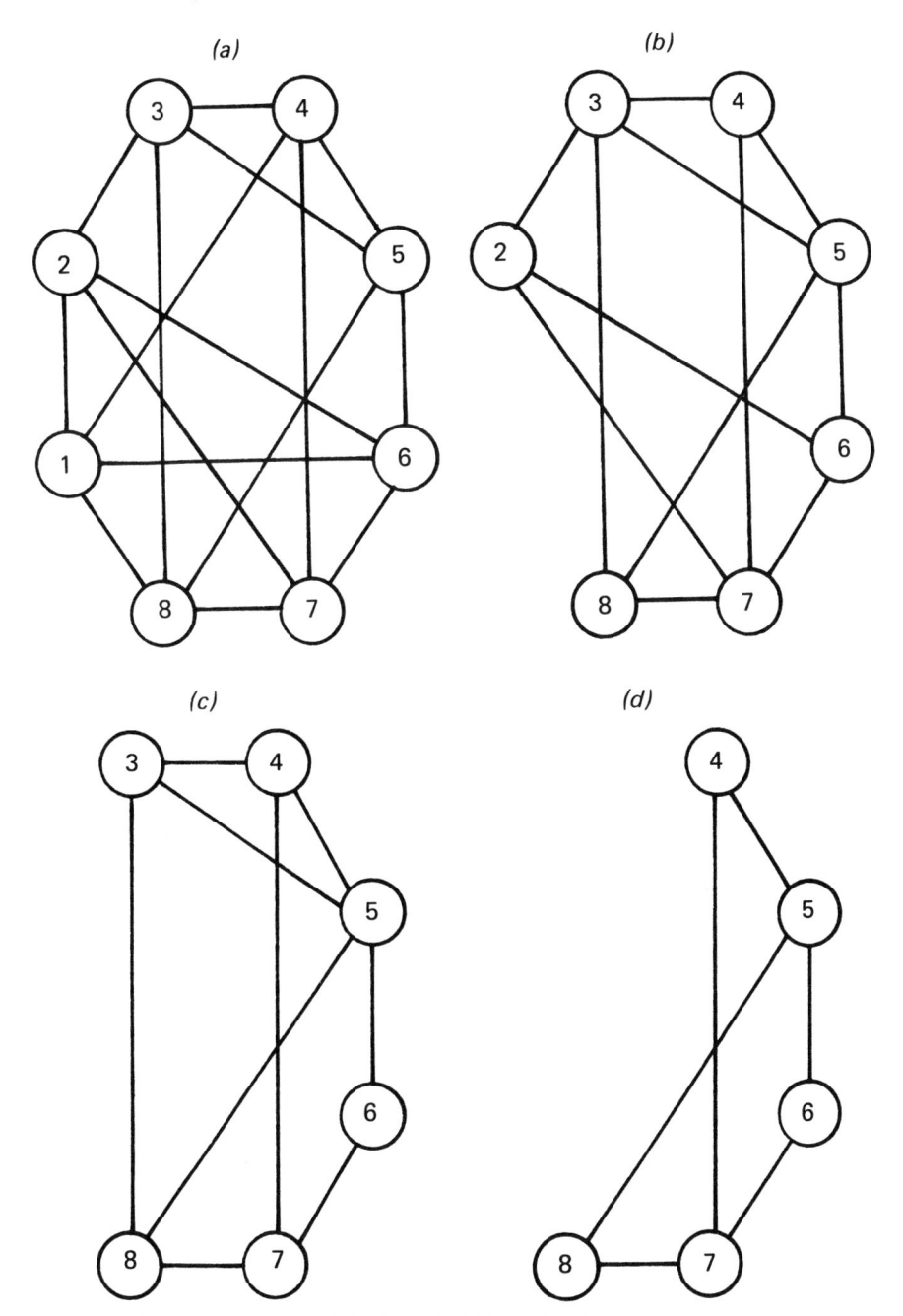

Figure 4-6. Network to Illustrate Kleitman's Theorem

Adapted from *"Network Analysis"* by H.Frank and I.T. Frisch. Copyright © 1970 by Scientific American, Inc. All rights reserved.

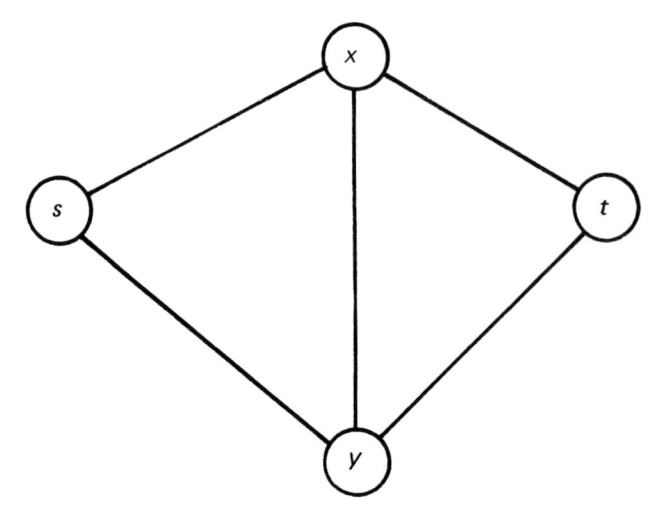

Figure 4-7. Network for Terminal Reliability Calculation

abilities are zero so that only link failures may occur. Denote q_{ij} the complementary probability $1 - p_{ij}$, that is, the probability that link (i,j) is operable. Next, number the routes from s to t from 1 to R, the total number of routes; the order is immaterial. For the example, we have:

Route 1 $(s,x), (x,t)$
Route 2 $(s,x), (x,y), (y,t)$
Route 3 $(s,y), (y,x), (x,t)$
Route 4 $(s,y), (y,t)$

The required probability Q_{st} can be found by a straightforward application of the rule for finding the probability that at least one among R events occurs [9]. Let E_i, $i = 1, \ldots, R$ be the event that route i between s and t is operable, $E_i E_j$ the joint event that routes i and j are both operable, and so forth for events $E_i E_j E_k$, and so on; in each case, we will assume $i < j < k \ldots \leq R$ so that each combination of routes has a unique representation. Similarly, let P_i be the probability of event E_i, P_{ij} the probability of event $E_i E_j$, and so on. If we now write $S_1 = \Sigma P_i$, $S_2 = \Sigma P_{ij}$, $S_3 = \Sigma P_{ijk}$, ..., we have

$$Q_{st} = S_1 - S_2 + S_3 - \cdots \pm S_R \qquad (4.3)$$

Each term S_m in equation 4.3 is the sum of the $R!/[m!(R-m)!]$ distinct probabilities for a given m routes out of R being operable. For example, in

Reliability of Networks

figure 4–7, $S_1 = P_1 + P_2 + P_3 + P_4$, $S_2 = P_{12} + P_{13} + P_{14} + P_{23} + P_{24} + P_{34}$, $S_3 = P_{123} + P_{124} + P_{134} + P_{234}$, and $S_4 = P_{1234}$.

To obtain the $P_i, P_{ij}, \ldots$, in terms of the link probabilities q_{ij}, we note first that P_k equals the product of the q_{ij} for route k. For example, in figure 4–7, $P_2 = q_{sx}q_{xy}q_{yt}$. However, it is not true in general that $P_{kl} = P_k P_l$, or similarly for P_{klm}, and so on. Thus in figure 4–7, $P_1 P_2 = q_{sx}^2 q_{xt} q_{xy} q_{yt}$. The term q_{sx} occurs to the second power because link (s, x) is common to routes 1 and 2. The correct value of P_{12} is $q_{sx} q_{xt} q_{xy} q_{yt}$ since it is the product of the q_{ij} for all links in either route 1 or route 2, each q_{ij} appearing only once. To get around this difficulty, we use a special multiplication rule developed by Lee [10]:

$$q_{ij}^a q_{kl}^b = q_{ij} \text{ if } (i,j) \text{ is the same link as } (k,l);$$
$$a \text{ and } b \text{ are any exponents} \geq 1 \quad (4.4)$$

We symbolize this by writing, for example, $P_1 * P_2$, which means that the product $P_1 P_2$ is formed by conventional multiplication, and then the rule of equation 4.4 is applied so that no q_{ij} occurs to a higher power than the first. Thus example, if each $q_{ij} = 0.9$, then the value of $P_1 P_2$ is $(0.9)^5 = 0.59049$, whereas that of $P_1 * P_2$ is $(0.9)^4 = 0.6561$. Note that $P_{ij} = P_i * P_j$, $P_{ijk} = P_i * P_j * P_k$, and so on.

The multiplication (*) operation can be carried out conveniently in a computer program as follows. First, number the links from 1 to M, where M is the total number of links. Next, represent each route between s and t by an M-place binary number in which the kth bit (that is, the bit whose value is 2^{k-1}, $k = 1, \ldots, M$) is 1 if link k is in the route and is zero if it is not. To illustrate, suppose the links in figure 4–7 are numbered as follows:

Link	Link Number
(s, x)	1
(s, y)	2
(x, t)	3
(x, y)	4
(y, t)	5.

Then the binary number representing route 1 is 00101 and that representing route 2 is 11001. The binary number "representing" $P_{12} = (q_{sx} q_{xt}) * (q_{sy} q_{yt})$ is found by taking the bit-by-bit logical sum (inclusive OR) of these two numbers; if we indicate such a sum by $\otimes$, this is $00101 \otimes 11001 = 11101$. Translating this result back into the links it represents, the (*) product is $q_{sx} q_{xt} q_{xy} q_{yt}$. The method applies to (*) products for any number of routes. Thus if binary number B_i represents the links in route i, then binary number

$B_m \oplus B_n \oplus \cdots \oplus B_x$ represents the represents the links in the (*) product $P_m * P_n * \cdots * P_x$.

It is difficult to apply this method directly to any but small networks. For one thing, we require a list of all R routes between s and t. As we have seen in chapter 3, R becomes quite large as the network size increases, and it is time-consuming even with a computer to generate this list. Also the number of distinct P terms that make up the right side of equation 4.3 is $2^R - 1$; even though the combinations of the $P_i, P_{ij}, \ldots$, can be represented by R-place binary numbers, somewhat as in the technique described earlier for finding (*) products, the amount of computation to find Q_{st} grows very rapidly with R. The situation is improved if we are interested only in the probability Q'_{st} that at least one of some smaller number $R' < R$ of routes is operable, for example, if we restrict the choices to a list of acceptable routes (see section 3-5).

The computational difficulty here has given rise to various alternative methods for finding the terminal reliability, to be discussed.

Methods Based on Link Cut-sets

In equation 4.3, Q_{st} is expressed as the probability that at least one route between s and t is operable. The complementary probability $1 - Q_{st}$, that no route is operable, is also the probability that all the links in at least one s, t link cut-set fail (see section 4-2.). In notation much like that used before, let E^i be the event that all links in s, t link cut-set i fail; $E^i E^j$ the joint event that all links in both s, t link cut-sets i and j fail, and so on; P^i equal the probability of event E^i; P^{ij} equal the probability of event $E^i E^j$, and so on; $S^1 = \Sigma P^i$, $S^2 = \Sigma P^{ij}$, and so on, with $i < j \le C$, the total number of s, t link cut sets. Then, just as before, we have

$$1 - Q_{st} = S^1 - S^2 + S^3 - \cdots \pm S^C \qquad (4.5)$$

In figure 4-7, for example, number the s, t link cut-sets arbitrarily:

Number	s, t Link Cut-set
1	$(s, x), (s, y)$
2	$(x, t), (y, t)$
3	$(s, y), (y, x), (x, t)$
4	$(s, x), (x, y), (y, t)$

Then $P^1 = p_{sx} p_{sy}$, $P^2 = p_{xt} p_{yt}$, and so on. The (*) multiplication rule, equation 4.4, now applies to the p_{ij} in finding $P^{1\,2} = P^1 * P^2$, and so on, just as before.

The motivation for expressing Q_{st} in terms of the s, t link cut-set failure probabilities is that when the average node degree $[2l/n]$ (section 4–2) exceeds 4, the number of s, t link cut-sets tends to be less than the number of routes between s and t [11]. Thus it is simpler to compute Q_{st} from equation 4.5 than from equation 4.3. A method of enumerating the s, t link cut-sets is given by Jensen and Bellmore [12].

Part of the remaining complexity in computing equation 4.5 arises from the mutual relationships among the events E^i, that is, among the failures of the s, t link cut-sets. Suppose that instead of events E^i, we deal with a set of events A^j with the following characteristics: (a) the occurrence of any one of the A^j renders all routes between s and t inoperable; (b) A^j and A^k are mutually exclusive, that is, the probability of $A^j A^k$ is zero for $j \neq k$; and (c) the A^j collectively exhaust all ways in which s and t may be disconnected. Then, if P_A^j is the probability of occurrence of A^j, we have $1 - Q_{st} = \Sigma P_A^j$; only the term corresponding to S^1 in equation 4.5 remains.

Hänsler et al. give an algorithm for finding a set of events A^j, each of which is actually a compound event $E^i B^{ij}$ in which E^i is the failure of s, t link cut-set i, and B^{ij} consists of the failed and operable states of other links [11]. Their method does not require a separate enumeration of the s, t link cut-sets, and the total number of events A^j is not much greater than the number of these cut-sets.

Node Failures

By applying the principle that the failure of a node implies the failure of all links incident at the node, Aggarwal et al. arrive at a simple method for finding Q_{st} in the presence of node failures, once an expression for it has been found without node failures, for example, by means of equation 4.3. [13] They define the modified link-operable probability, for a link in an undirected graph, as $q'_{ij} = q_i q_j q_{ij}$, where $q_i = 1 - p_i$ is the node-operable probability. In figure 4–7, for example, $q_x = q_s q_x q_{sx}$, and so on. These modified link probabilities are simply substituted for the original ones in evaluating Q_{st}; the (∗) multiplication rule of equation 4.4 is applied not only to the q_{ij} but also to the node probabilities q_i. Thus in the example, $P_{1'2'} = P_{\cdot} * P_{2'} = (q'_{sx} q'_{sy}) * (q'_{xt} q'_{yt}) = (q_s q_x q_{sx} q_s q_y q_{sy}) * (q_x q_t q_{xt} q_y q_t q_{yt}) = q_s q_x q_{sx} q_y q_{sy} q_t q_{xt} q_{yt}.$

References

1. Wilkov, R.S. "Analysis and Design of Reliable Computer Networks." *IEEE Trans. Commun.* COM-20:660–678, 1972.

2. Ford, L.R., Jr., and Fulkerson, D.R. *Flows in Networks*. Princeton, N.J.: Princeton University Press, 1962.
3. Frank, H., and Frisch, I.T. *Communication, Transmission, and Transportation Networks*. Reading, Mass.: Addison-Wesley, 1971.
4. Frank, H., and Frisch, I.T., "Analysis and Design of Survivable Networks," *IEEE Trans. Commun. Tech.* COM-18:501–519, 1970.
5. Price, W.L. *Graphs and Networks, an Introduction*. London: Butterworth, 1971.
6. Gomory, R.E., and Hu, T.C. "Multiterminal Network Flows," *SIAM J. Appl. Math.* 9:551–570, 1971.
7. Kleitman, D.J. "Methods for Investigating Connectivity of Large Graphs," *IEEE Trans. Circuit Theory* CT–16:232–233, 1969.
8. Frank, H., and Frisch, I.T. "Network Analysis," *Scientific American* 223:94–103, July 1970.
9. Feller, W. *An Introduction to Probability Theory and Its Applications*. Vol. I. 2d ed. New York: Wiley, 1957.
10. Lee, C.Y. "Analysis of Switching Networks," *Bell Syst. Tech. J.* 34:1287–1315, 1955.
11. Hänsler, E., et al. "Exact Computation of Computer Network Reliability," *Networks* 4:95–112, 1974.
12. Jensen, P.A., and Bellmore, M. "An Algorithm to Determine the Reliability of a Complex System," *IEEE Trans. Reliability* R-18:169–174, 1969.
13. Aggarwal, K.K. et al. "A Simple Method for Reliability Evaluation of a Communication System." *IEEE Trans. Commun.* COM–23:563–566, 1975.

**Part II
Switched Networks**

5 Circuit-Switched Networks

5-1 Introduction

In a circuit-switched network, a temporary communications path is set up from one node, or switching point, to another, when it is required by a pair of users. This path may traverse intermediate switching points. The users may communicate over the path as soon as it is established; when the communications session is over, the switching equipment breaks down the path, and the portions of the network links and the switching equipment that were in use are once again available to other users.

The most familiar example is the telephone system, in which each subscriber is connected to a switching point, the *local central office*, by a permanent link, the *subscriber loop*. Central offices are connected by *trunk groups*, which are links with capacities of one circuit to several hundred circuits, or, in long-haul systems, up to several thousand circuits. In the case of a call between two directly connected central offices, the switching equipment in the one from which the call originates selects an idle circuit and controls the setup of the coversation path to the switching equipment in the called central office; the latter equipment extends the path to the called subscriber. In more complex networks, some of the switching functions are carried out by equipment at intermediate nodes.

Circuit switching makes it possible to use the trunk groups and switching equipment economically since subscribers share these facilities and use them only during their conversations. Suppose, for example, that there are 5,000 subscribers connected to one central office and 3,000 to another, nearby. In normal circumstances, it is improbable even during the busiest part of the day that more than, say, 50 conversation paths will be needed simultaneously between the two groups of subscribers. Thus if there is a trunk group of capacity 50 circuits between the central offices, only rarely will the demand exceed this number, with the result that calls by additional subscribers are *blocked*, that is, unable to find an idle circuit. The same principle extends to complex networks in which there are many central offices and in which there must be at least one idle circuit in each of several trunk groups (links) for a path to be set up: the capacities of the trunk groups are based on having only a modest proportion, usually a few percent or less, of call attempts blocked during the busiest period of the day.

Traffic engineering is the application of probability theory to the problem of estimating trunk-group sizes so that, on the one hand, users will receive satisfactory service, and on the other hand, the telephone company or private system owner will not overinvest in facilities. In this chapter, we deal with the methods of traffic engineering that are of most use in the analysis of circuit-switched networks, progressing in order of complexity from single trunk groups, to multiple groups with alternate routing, to complete networks.

5-2 The Full-availability Trunk Group

The simplest circuit-switched network consists of two nodes, or switching points, connected by one link, or trunk group. We will show how the capacity of this trunk group can be estimated so that, for certain assumed characteristics of the traffic, the probability of call blocking will not exceed a prescribed value.

Basic Terms and Assumptions

We idealize the behavior of subscribers and the operation of the switching equipment in certain aspects. We assume that the switching equipment can select any idle circuit in the group to serve any call request; this is the assumption of *full availability* of the trunk group. Further, we assume that if all circuits are busy when there is a new call request, the equipment sends the subscriber a *busy signal* and his call request is immediately *cleared* from the system. This potential call is said to have been *blocked* or *lost*.

Another important assumption is that subscribers make call requests at random, or in more precise terms, that the input is *Poissonian*. In other words, the probability $p_j(t)$ that exactly j call requests will occur in an interval of t seconds is

$$p_j(t) = \frac{(\lambda t)^j \exp(-\lambda t)}{j!} \quad j = 0, 1, 2, \ldots \quad (5.1)$$

In equation 5.1, λ is a positive constant with the following meaning: the intervals between successive call requests, that is, the *interarrival times*, have a negative exponential distribution with mean $1/\lambda$ seconds. Thus the probability that an interarrival time is less than or equal to t_1 seconds is $1 - \exp(-\lambda t_1)$, $t_1 \geq 0$. Further, the average rate of call arrivals is λ per second.

The assumption of random call requests embodied in equation 5.1 can be strictly true only when the number of subscribers who make call requests is infinite, a condition that appears to depart from reality. However, as long as the number of subscribers is several times larger than the number of circuits

Circuit-Switched Networks

in the trunk group, which is true in most situations, the Poisson assumption is not unreasonable.

We assume that the average call duration, or *holding time*, is $1/\mu$ seconds. We omit the assumption made in some developments of traffic theory that the holding times have a negative exponential distribution since the Erlang B formula applies for an arbitrary distribution [1].

We assume that the parameters λ and μ that characterize the statistics of the call requests and call durations are constants, not dependent on the time. Furthermore, we assume that the trunk group is in statistical equilibrium, which we can imagine has been attained by starting with all circuits idle and allowing calls to be served for an interval that is at least several times the average holding time, $1/\mu$. It is customary in practice to focus attention on the busy hour, that is, a period during which call activity is at its greatest relative to other periods, and which has steadystate characteristics. (This period need not be 60 minutes long, and perhaps should be called the *busy period*. Traffic engineers may refer to the "9 a.m. to 10:30 a.m. busy hour.")

The Erlang B Formula

When the assumptions of full availability, clearing of lost calls, Poissonian input, and statistical equilibrium are met for a trunk group, the probability that a call is lost is given by the *Erlang B formula*:

$$\Pr(\text{lost call}) = B(c, a) = \frac{\left(\dfrac{a^c}{c!}\right)}{\sum_{i=0}^{c}\left(\dfrac{a^i}{i!}\right)} \qquad (5.2)$$

In equation 5.2, c is the number of circuits in the trunk group, and the dimensionless quantity $a = \lambda/\mu$ is called the *offered traffic in erlangs*. (The word "offered" is used to describe the traffic or usage that would be generated by call requests if there were no loss of calls. The contrary term "carried" applies to traffic or usage that actually appears on the circuits. The unit erlang is named for the Danish traffic theorist, A.K. Erlang.) In practice, the call-arrival rate λ is estimated by observing or assuming that there are N call requests in a period of T seconds, say, in the busy hour; then $\lambda = N/T$. This is combined with an observed or assumed value of $1/\mu$, the average holding time, to calculate a, the offered traffic. Suppose, for example, that subscribers make 500 call requests between 9:00 a.m. and 10:15 a.m., and that the average holding time is 200 seconds. Then we have $a = (500 \times 200)/(75 \times 60) = 22.2$ erlangs of offered traffic. Equation 5.2

shows that with a trunk group of $c = 29$ circuits, a proportion $B(29, 22.2) = 0.0312$ of the calls is lost, whereas with $c = 34$ circuits, a proportion $B(34, 22.2) = 0.00480$ of the calls is lost. The proportion B is referred to as the *grade of service*.

As the number of circuits c in the trunk group becomes indefinitely large, the loss probability $B(c, a)$ tends toward zero. Suppose we were to observe in this case, in statistical equilibrium, the instantaneous number of occupied circuits, $b(t)$, for a period of time $t_1 \leq t \leq t_2$, with $t_2 - t_1 \gg 1/\mu$. At any instant, $b(t)$ could have any nonnegative integer value. [With Poissonian input, $b(t)$ cannot increase or decrease by more than 1 at any instant.] Further, the average value of $b(t)$, $\bar{b} = (t_2 - t_1)^{-1} \int_{t_1}^{t_2} b(u)du$ would tend, as $t_2 - t_1 \to \infty$, toward a, the offered traffic. This statement is actually the basis for the definition of the traffic in erlangs, viz., the total occupancy or usage of the trunk group during a period $t_1 \leq t \leq t_2$, divided by the length of the period, $t_2 - t_1$. On the other hand, if we were to make similar observations of the instantaneous number of occupied circuits in the actual finite trunk group of c circuits, we would have $0 \leq b(t) \leq c$, and we would find a smaller average value b, owing to the loss of some calls. In fact, this latter value would tend toward the *carried traffic*, a':

$$a' = a[1 - B(c, a)] \qquad (5.3)$$

In the preceding example, with $c = 29$ circuits, we have $a' = 22.2 \times (1 - 0.0312) = 21.5$ erlangs. Equation 5.3 is useful when an estimate of a must be made by observing the trunk group and finding a'. {Practical traffic-measuring instruments obtain an estimate of a' by scanning all c circuits periodically, thus obtaining a series of sample values of $b(t)$. This is done for an interval that is long compared to the average holding time, say 30 minutes to 1 hour for telephone calls. The error that is introduced by sampling is small if the sampling period is substantially less than the average holding time, say, 10 to 100 seconds for telephone calls [2].} We note that the blocked traffic, that is, the difference $a - a'$, is, from equation 5.3, $aB(c, a)$.

The Erlang B loss function, as in equation 5.2, has been tabulated for ranges of values of the number of circuits c and offered traffic a [3]. A table in which the values of a are given for fixed c and probability of loss is called a *capacity table*, as in table 5–1. This table shows, for example, that a trunk group of 50 circuits with 37.9 erlangs of offered traffic will have a loss probability of 0.01. As a further example, if we have 11.1 erlangs of offered traffic, we will need 20 circuits for a loss probability of 0.005. By means of linear interpolation, table 5–1 can be used to estimate $B(c, a)$ for values of c or a other than those in the table. The simplest way to calculate $B(c, a)$ in a computer program is with the recurrence relation

Table 5-1
Erlang B Capacity Table
(Offered Traffic, a, for Indicated Blocking Probability)

Number of Circuits, c	0.001	0.002	0.005	0.01	0.02	0.03	0.05	0.1	0.2
1	0.001	0.002	0.005	0.010	0.020	0.031	0.053	0.111	0.250
2	0.046	0.065	0.105	0.153	0.223	0.282	0.381	0.595	1.00
3	0.194	0.249	0.349	0.455	0.602	0.715	0.899	1.27	1.93
4	0.439	0.535	0.701	0.869	1.09	1.26	1.52	2.05	2.95
5	0.762	0.900	1.13	1.36	1.66	1.88	2.22	2.88	4.01
6	1.15	1.33	1.62	1.91	2.28	2.54	2.96	3.76	5.11
7	1.58	1.80	2.16	2.50	2.94	3.25	3.74	4.67	6.23
8	2.05	2.31	2.73	3.13	3.63	3.99	4.54	5.60	7.37
9	2.56	2.85	3.33	3.78	4.34	4.75	5.37	6.55	8.52
10	3.09	3.43	3.96	4.46	5.08	5.53	6.22	7.51	9.68
11	3.65	4.02	4.61	5.16	5.84	6.33	7.08	8.49	10.9
12	4.23	4.64	5.28	5.88	6.61	7.14	7.95	9.47	12.0
13	4.83	5.27	5.96	6.61	7.40	7.97	8.83	10.5	13.2
14	5.45	5.92	6.66	7.35	8.20	8.80	9.73	11.5	14.4
15	6.08	6.58	7.38	8.11	9.01	9.65	10.6	12.5	15.6
16	6.72	7.26	8.10	8.88	9.83	10.5	11.5	13.5	16.8
17	7.37	7.95	8.83	9.65	10.7	11.4	12.5	14.5	18.0
18	8.05	8.64	9.58	10.4	11.5	12.2	13.4	15.5	19.2
19	8.72	9.35	10.3	11.2	12.3	13.1	14.3	16.6	20.4
20	9.41	10.1	11.1	12.0	13.2	14.0	15.2	17.6	21.6
21	10.1	10.8	11.9	12.8	14.0	14.9	16.2	18.7	22.8
22	10.8	11.5	12.6	13.7	14.9	15.8	17.1	19.7	24.1
23	11.5	12.3	13.4	14.5	15.8	16.7	18.1	20.7	25.3
24	12.2	13.0	14.2	15.3	16.6	17.6	19.0	21.8	26.5
25	13.0	13.8	15.0	16.1	17.5	18.5	20.0	22.8	27.7
26	13.7	14.5	15.8	17.0	18.4	19.4	20.9	23.9	28.9
27	14.4	15.3	16.6	17.8	19.3	20.3	21.9	24.9	30.2

Table 5-1 *(continued)*

Number of Circuits, c	0.001	0.002	0.005	0.01	0.02	0.03	0.05	0.1	0.2
28	15.2	16.1	17.4	18.6	20.2	21.2	22.9	26.0	31.4
29	15.9	16.8	18.2	19.5	21.0	22.1	23.8	27.1	32.6
30	16.7	17.6	19.0	20.3	21.9	23.1	24.8	28.1	33.8
31	17.4	18.4	19.9	21.2	22.8	24.0	25.8	29.2	35.1
32	18.2	19.2	20.7	22.0	23.7	24.9	26.7	30.2	36.3
33	19.0	20.0	21.5	22.9	24.6	25.8	27.7	31.3	37.5
34	19.7	20.8	22.3	23.8	25.5	26.8	28.7	32.4	38.8
35	20.5	21.6	23.2	24.6	26.4	27.7	29.7	33.4	40.0
36	21.3	22.4	24.0	25.5	27.3	28.6	30.7	34.5	41.2
37	22.1	23.2	24.8	26.4	28.3	29.6	31.6	35.6	42.4
38	22.9	24.0	25.7	27.3	29.2	30.5	32.6	36.6	43.7
39	23.7	24.8	26.5	28.1	30.1	31.5	33.6	37.7	44.9
40	24.4	25.6	27.4	29.0	31.0	32.4	34.6	38.8	46.1
41	25.2	26.4	28.2	29.9	31.9	33.4	35.6	39.9	47.4
42	26.0	27.2	29.1	30.8	32.8	34.3	36.6	40.9	48.6
43	26.8	28.1	29.9	31.7	33.8	35.3	37.6	42.0	49.9
44	27.6	28.9	30.8	32.5	34.7	36.2	38.6	43.1	51.1
45	28.4	29.7	31.7	33.4	35.6	37.2	39.6	44.2	52.3
46	29.3	30.5	32.5	34.3	36.5	38.1	40.5	45.2	53.6
47	30.1	31.4	33.4	35.2	37.5	39.1	41.5	46.3	54.8
48	30.9	32.2	34.2	36.1	38.4	40.0	42.5	47.4	56.0
49	31.7	33.0	35.1	37.0	39.3	41.0	43.5	48.5	57.3
50	32.5	33.9	36.0	37.9	40.3	41.9	44.5	49.6	58.5

Circuit-Switched Networks

55	36.6	38.1	40.4	42.4	44.9	46.7	49.5	55.0	64.7
60	40.8	42.4	44.8	46.9	49.6	51.6	54.6	60.4	70.9
65	45.0	46.6	49.2	51.5	54.4	56.4	59.6	65.8	77.1
70	49.2	51.0	53.7	56.1	59.1	61.3	64.7	71.3	83.3
75	53.5	55.3	58.2	60.7	63.9	66.2	69.7	76.7	89.5
80	57.8	59.7	62.7	65.4	68.7	71.1	74.8	82.2	95.7
85	62.1	64.1	67.2	70.0	73.5	76.0	79.9	87.7	102.0
90	66.5	68.6	71.8	74.7	78.3	80.9	85.0	93.1	108.2
95	70.9	73.0	76.3	79.4	83.1	85.8	90.1	98.6	114.4
100	75.2	77.5	80.9	84.1	88.0	90.8	95.2	104.1	120.6
110	84.1	86.4	90.1	93.5	97.7	100.7	105.5	115.1	133.1
120	93.0	95.5	99.4	103.0	107.4	110.7	115.8	126.1	145.6
130	101.9	104.6	108.7	112.5	117.2	120.6	126.1	137.1	158.0
140	110.9	113.7	118.0	122.0	127.0	130.6	136.4	148.1	170.5
150	119.9	122.9	127.4	131.6	136.8	140.6	146.7	159.1	183.0
160	129.0	132.1	136.8	141.2	146.6	150.6	157.0	170.2	195.5
170	138.1	141.3	146.2	150.8	156.5	160.7	167.4	181.2	207.9
180	147.3	150.6	155.7	160.4	166.4	170.7	177.8	192.2	220.4
190	156.4	159.8	165.2	170.1	176.3	180.8	188.1	203.3	232.9
200	165.6	169.2	174.6	179.7	186.2	190.9	198.5	214.3	245.4

$$B(k, a) = \frac{aB(k-1, a)}{[k + aB(k-1, a)]} \quad k = 1, 2, 3, \ldots, c \quad (5.4)$$

starting with $B(0, a) = 1$. Numerical methods for approximating $B(c, a)$, particularly for large values of c, are given by Jagerman [4].

Some tables use a different unit of traffic, called *CCS*, an abbreviation for "hundreds of call-seconds per hour." In the earlier example with 500 potential calls of average duration 200 seconds made in 1.25 hours, we have $(500 \times 200)/(100 \times 1.25) = 800$ CCS of offered traffic. In general, if N call requests are made in T seconds, and the average holding time is $1/\mu$ seconds, we have $(N/100\ \mu)/(T/3,600)$ CCS; thus the traffic in CCS is 36 times its value in erlangs.

5-3 The Trunk Group under Other Assumptions

If any of the assumptions of section 5-2 is not met, the Erlang B formula cannot be used to estimate the probability of loss. Here we discuss some of the more important departures from these assumptions.

Partial Availability

Some types of switching equipment may not be able to select any one of the c circuits in the trunk group but rather may be limited to some number less than c. This is called *partial availability*. For example, table 5-1 shows that with full availability, 30 circuits can handle 21.9 erlangs of offered traffic with a grade of service of 0.02. Suppose, though, that the availability is limited to only 10 circuits, and the group of 30 circuits is operated as three separate subgroups of 10 circuits each. The capacity of each subgroup of 10 circuits, again referring to table 5-1, is 5.08 erlangs; thus the entire group of 30 circuits can handle only $3 \times 5.08 = 15.2$ erlangs, or 31.5 percent less than at full availability. The difference between these two situations is shown schematically in figure 5-1, in a form that is often used to diagram trunking arrangements; in practice, a scheme such as this is realized by the switching equipment giving only one third of the subscribers access to circuits 1 through 10, and so on. This example illustrates the important principle that small groups of circuits are less efficient than large groups.

The example of figure 5-1 is the simplest type of *grading* or subdivision of the circuits in a trunk group with partial availability. More complex gradings, in which circuits are available to two or more parts of the total offered traffic, are discussed in the literature of traffic engineering [1, 5].

Circuit-Switched Networks

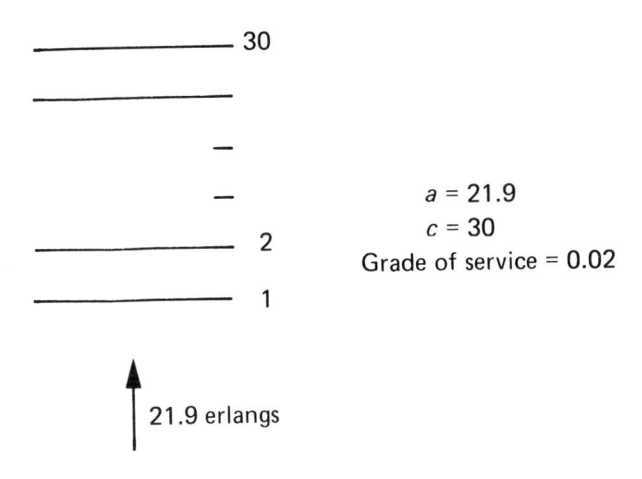

```
_____ 30
_____
    —
    —
_____ 2       a = 21.9
_____ 1       c = 30
                    Grade of service = 0.02

  ↑
  | 21.9 erlangs
```

(a) Full availability

Grade of service = 0.02 in each subgroup

```
_____ 10        _____ 20        _____ 30
   —                —                —
   —                —                —

_____ 1         _____ 11        _____ 21
  ↑                ↑                ↑
  |                |                |

a₁ = 5.08       a₂ = 5.08       a₃ = 5.08
```

$a_1 = 5.08 \qquad a_2 = 5.08 \qquad a_3 = 5.08$

(b) Partial availability

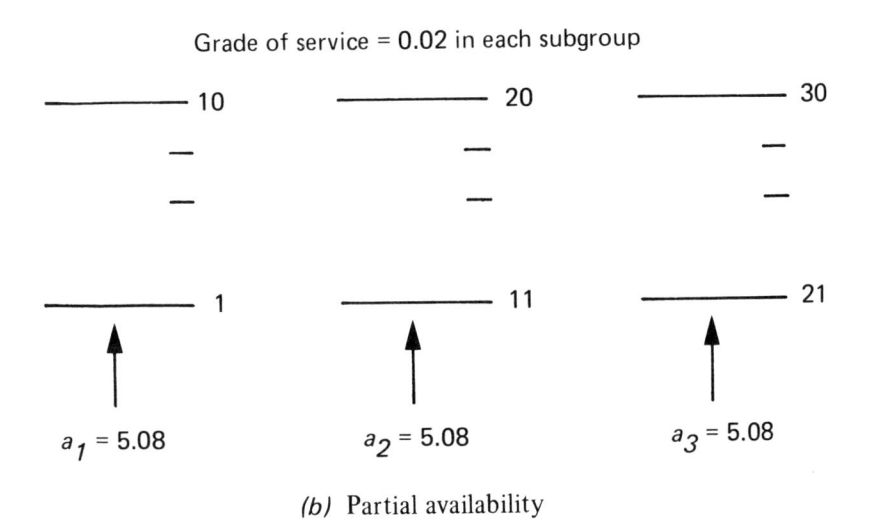

Figure 5–1. Full versus Partial Availability

Nonrandom Call Requests

The pattern of call requests may depart from randomness in any of several ways. For example, we have noted that the assumption of random or Poissonian input expressed in equation 5.1 can hold only if the number of subscribers is infinite or, in practical terms, when it is much larger than c, the number of circuits in the trunk group. When the number of subscribers n is only moderately greater than c, the call-blocking probability with full availability is given by the *Engset* formula. We imagine here that the n subscribers have access to the trunk group at only one end, that is, only for calls that they originate to a population of subscribers at the other end that is much more numerous. This is the case, for example, when n subscribers at a private branch exchange (PBX) make outgoing calls over c exchange circuits.

$$\text{Pr(lost call)} = P(c, n, b') = \frac{\binom{n-1}{c}(b')^c}{\sum_{i=0}^{c}\binom{n-1}{i}(b')^i} \quad (5.5)$$

Here b' is given by

$$b' = \frac{b}{1 - b(1 - P)} \quad (5.6)$$

in which b is the offered traffic per subscriber and P is the probability given by equation 5.5. It is because of the dependence of b' on P, which necessitates an iterative calculation of P, and the complexity of equation 5.5 that the Engset formula is seldom used. When it is used, the blocking probability $P(c, n, b')$ for a given total offered traffic and number of circuits c is slightly lower than that given by the Erlang B formula since the presence of $k \leq c$ calls, in the Engset case, means that k of the n subscribers are occupied, and it is thus less likely that further call requests will be made. For example, suppose we have $n = 25$ subscribers and $b = 0.1$ erlangs per subscriber, or $a = 2.5$ erlangs offered altogether. A table of the Engset loss probability shows that the proportion of calls blocked with $c = 6$ circuits is 0.0204, whereas with the Erlang B formula we find $B(6, 2.5) = 0.0282$ [6].

Another kind of departure from random call requests occurs because of the behavior of users. A telephone subscriber who encounters blocking is likely to make further requests for the same call in rapid succession until his

Circuit-Switched Networks

call is connected or he becomes discouraged and abandons it. In switched data-communications systems, automatic equipment may generate repeated requests, or they may be made manually at terminals. In extreme cases, when repeated requests are made on many of the user lines, it can cause serious congestion of the switching equipment because this equipment is engineered not only for a prescribed amount of offered traffic in the busy hour but also for a prescribed number of call requests, or attempts, in this period.

An approximate method to account for repeated call attempts is as follows. Suppose that the original, or first attempt, offered traffic is a, and that, owing to repeated attempts, an added amount of traffic Δa is offered. Thus the total offered traffic in the busy hour is $a_R = a + \Delta a$. If we assume that the totality of call requests, both original and repeated, is still Poissonian,[a] then the blocking probability is $B(c, a_R)$, and the blocked traffic is $a_R B(c, a_R)$. If we now assume that Δa consists of a proportion $\rho (0 \leq \rho \leq 1)$ of this blocked traffic, then we have for the total offered traffic

$$a_R = a + \rho \, a_R B(c, a_R) \tag{5.7}$$

For given values of a, c, and ρ, equation 5.7 can be solved recursively for a_R. For example, suppose that $a = 4.0$ erlangs, $c = 6$ circuits, and $\rho = 0.5$; that is, half the blocked traffic is reoffered. Then we must find a_R as the solution of

$$a_R = 4.0 + 0.5 a_R B(6, a_R)$$

This can be done, for example, by setting $F(a_R) = 4.0 + 0.5 a_R B(6, a_R)$ and finding a solution of the equation $F(a_R) = a_R$ by the method of successive substitution, starting with the initial value $a_R \approx a = 4.0$:

$$F(4.0) = 4.0 + [0.5 \times 4.0 B(6, 4.0)] = 4.24$$
$$F(4.24) = 4.0 + [0.5 \times 4.24 B(6, 4.24)] = 4.29$$
$$F(4.29) = 4.0 + [0.5 \times 4.29 B(6, 4.29)] = 4.30$$
$$F(4.30) = 4.0 + [0.5 \times 4.30 B(6, 4.30)] = 4.30$$

Thus to three significant figures, $a_R = 4.30$ erlangs. The blocking probability is $B(6, 4.30) = 0.139$, and the carried traffic is, using equation 5.3, $4.30(1 - 0.139) = 3.70$ erlangs. In contrast, if we had not used equation 5.7 to account for the increase in offered traffic brought about by repeated attempts (or equivalently, if we had put $\rho = 0$), we would have found the

---

[a] Herein lies the approximate nature of the method. The combination of original and repeated call requests will in general not be Poissonian since the arrival instants of repeat requests are statistically correlated with those of the original attempts. The repeated-attempt problem has been solved without making this sort of approximation [1], but the solution is quite complex and has not been applied in practice.

blocking probability to be $B(6, 4.0) = 0.128$) and the carried traffic to be $4.0 \times (1 - 0.128) = 3.49$ erlangs. This conforms with our expectation: without repeated attempts, a smaller proportion of the original offered traffic is carried, but the blocking probability is lower.

If in equation 5.7 $\rho \approx 0$, or if the original blocking probability $B(c, a)$ is quite small, the effect of repeated attempts is negligible. If, on the other hand, $\rho = 1$, then equation 5.7 shows that $a_R[1 - B(c, a_R)] = a$, and *all* the original offered traffic a is carried, albeit with more blocking than originally (cf. equation 5.3). In our example, with $a = 4.0$, $c = 6$, and $\rho = 1$, $a_R = 4.91$ erlangs, and the blocking probability is $B(6, 4.91) = 0.19$. Wilkinson suggested that, for highly loaded trunk groups, a_R be estimated using an assumption equivalent to $\rho = 1$ [7]. More recently Jewett suggested that $\rho = 0.55$ is typical of telephone users whose attempts are first offered to a group of WATS circuits and then overflow to direct-distance-dialing (DDD) circuits [8].

5-4 Alternate Routing

Suppose in figure 5-2 that traffic between node 1 and node 2 is routed over the group of circuits that constitutes link (1, 2). It greatly increases the network reliability, that is, the probability of successful connection, if, when link (1, 2) is inoperable, calls from 1 to 2 can use one of the alternate routes {(1, 3), (3, 2)} or {(1, 4), (4, 5), (5, 2)} or {(1, 3), (3, 4), (4, 5), (5, 2)}. This illustrates one reason that alternate routing is widely used in circuit-switched networks. Another reason, not so obvious, is that of economy. Suppose that we wish to limit the blocking probability for calls from node 1 to node 2 to 2 percent. Without alternate routing, there must be enough circuits in group (1, 2) to achieve this level of blocking. If, in contrast, there is alternate routing, we can install fewer circuits in group (1, 2) and operate it at a much higher blocking probability, say, 10 percent. If the calls from node 1 to node 2 that are thus blocked can find an alternate route with a 20-percent chance of blocking, the users will perceive the desired 2-percent overall rate of blocking since only a proportion $0.1 \times 0.2 = 0.02$ of their calls will ultimately be blocked.[b] Thus savings are realized in trunk group (1, 2). Furthermore, links such as (1, 3) and (3, 2) in figure 5-2 will tend to carry both first-choice traffic, for example, from 1 to 3, as well as *overflow* or alternate-routed traffic, for example, from 1 to 2 via (1, 3) and (3, 2). Since there are generally economies of scale in trunk groups, that is, since large groups tend to cost less per circuit-mile that small ones, trunk groups such as (1, 3), (3, 2)

---

[b]Modern switches test the circuits in the direct route and in a number of alternate routes so rapidly that users are seldom aware of any delay in the process.

Circuit-Switched Networks

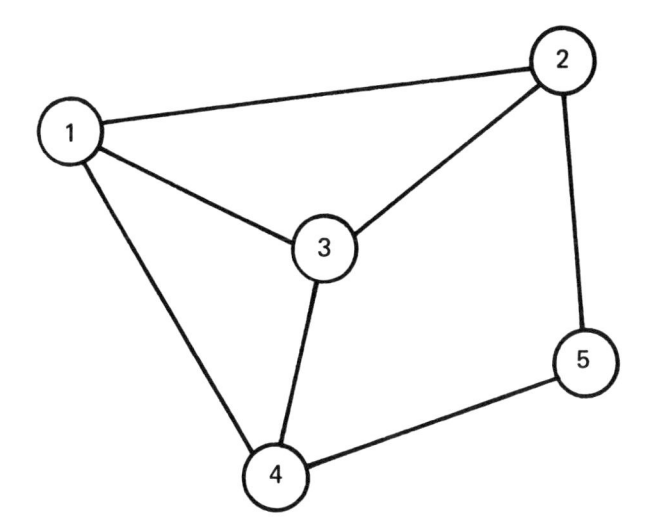

Figure 5-2. Network to Illustrate Alternate Routing

and (3, 4) in figure 5-2 will tend to be *backbone* groups, that is, large-capacity groups that carry diverse "parcels" of traffic, and thus further economies of scale result.

Peakedness of Overflow Traffic

Figure 5-3 shows the variation in time of the instantaneous number of busy circuits in a trunk group of 14 circuits; this hypothetical plot is for statistical equilibrium in the busy hour. We may get an intuitive idea of the *peakedness* of overflow traffic by looking, in figure 5-3, only at that part of the plot that lies above a horizontal line corresponding to some smaller number of circuits, say, 10. For example, between the times t_1 and t_2 in the figure, more than 10 circuits are busy. Thus if the trunk group contained only 10 circuits, the traffic represented by the shaded parts of the plot would be blocked, or, assuming that an alternate route were available, it would overflow to this route. The characteristic of the overflow traffic that distinguishes it from the original random offered traffic is that it consists of peaks separated by intervals of no traffic (see the shaded areas in figure 5-3).

In section 5-2 we noted that, when Poissonian traffic is offered to a group of circuits that is indefinitely large, that is, so large that the blocking probability is negligible, the average number of busy circuits is just a, the offered traffic in erlangs. (As noted in section 5-2, this statement is the basis for the definition of a in erlangs.) The probability distribution of the number of busy circuits in this situation is also Poissonian:

$$\Pr(x \text{ circuits busy}) = p(x) = \frac{a^x \exp(-a)}{x!} \quad x = 0, 1, 2, \ldots$$
(5.8)

The average number of busy circuits is, of course, $\Sigma_0^\infty u p(u) = a$. The variance of the number of busy circuits, viz., $\Sigma_0^\infty (u - a)^2 p(u)$, turns out also to have the value a. This is a distinctive aspect of random offered traffic: the *peakedness ratio*, defined as the ratio of the variance to the average, is 1. We might suspect from figure 5-3 that the peakedness ratio for overflow traffic is greater than 1. That this is the case is confirmed by the theoretical analysis of the probability distribution for the overflow traffic. We know from section 5-2 that the average value of the overflow (blocked) traffic when a erlangs of

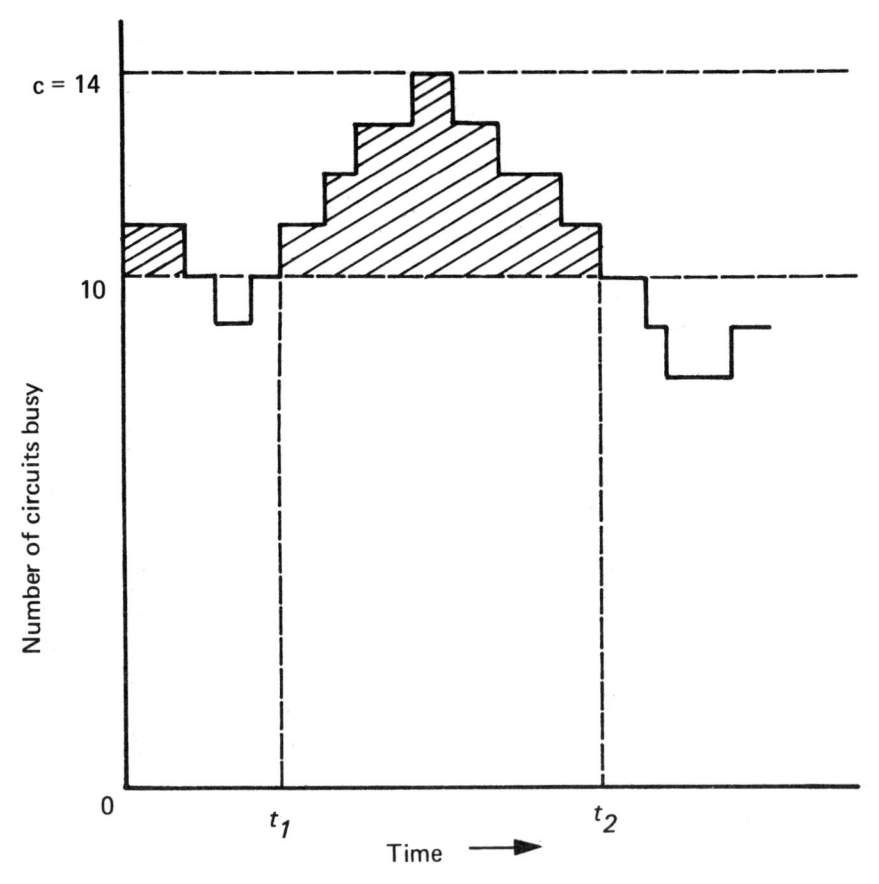

Figure 5-3. Number of Circuits Busy versus Time for a Trunk Group of 14 Circuits

Circuit-Switched Networks

random traffic is offered to c circuits is $aB(c, a)$. Just as in the case of random traffic, this is the average number of busy circuits when this overflow traffic is offered to a trunk group of indefinitely large size. An estimate of the variance of the overflow traffic is

$$v = b \left[\frac{1 - b + a}{c + 1 + b - a} \right] \quad (5.9)$$

in which b is the average value, that is $b = aB(c, a)$ [7]. For example, suppose that an amount $a = 5.53$ erlangs of random traffic is offered to a group of $c = 10$ trunks. Table 5-1 shows that the blocking probability is $B(10, 5.53) = 0.03$, and thus $b = 5.53 \times 0.03 = 0.166$ erlangs. The variance of this overflow traffic, from equation 5.9, is $v = 0.166[(1 - 0.166 + 5.53)/(10 + 1 + 0.166 - 5.53)] = 0.301$. The peakedness ratio is $v/b = 0.301/0.166 = 1.88$.

Blocking Probabilities with Alternate Routing

We cannot use the Erlang B formula, equation 5.2, to find the blocking probability when overflow traffic of average value a erlangs is offered to an alternate route group of c trunks because the formula applies only when the offered traffic is random, with peakedness ratio 1. The blocking probability for a given average value of traffic and number of circuits is greater for overflow traffic than for random traffic, owing to the greater peakedness of the former. Trunk groups carrying overflow traffic can be engineered by a method recommended by the CCITT, as follows [9].

The peakedness ratio of the overflow traffic, v/b, has a broad maximum as the random offered traffic a is varied with a fixed number of circuits c. Thus no consequential error is introduced by substituting for the actual peakedness ratio its maximum value, calculated by computer methods and given in table 5-2. For example, we found earlier that the peakedness ratio for the overflow traffic from $c = 10$ trunks with $a = 5.53$ erlangs of random traffic is 1.88; table 5-2 shows that the maximum peakedness ratio for $c = 10$ is 2.05.

Using the maximum peakedness ratio from table 5-2, we next consult table 5-3, which gives the number of circuits required for a blocking probability of 0.01 as a function of the average overflow traffic and the maximum peakedness ratio.[c] Suppose that, as in our earlier example, we have $a = 5.53$ erlangs of random traffic offered to $c = 10$ trunks, and let us use the

---

[c]Similar tables have been calculated for other blocking probabilities. The term "weighted mean peakedness ratio" in table 5-3 is explained later.

Table 5–2
Maximum Peakedness Ratio versus Number of Circuits

Number of Circuits	Maximum Peakedness Ratio
1	1.17
2	1.31
3	1.43
4	1.54
5	1.64
6	1.73
7	1.82
8	1.90
9	1.98
10	2.05
11	2.12
12	2.19
13	2.26
14	2.32
15	2.38
16	2.44
17	2.49
18	2.55
19	2.61
20	2.66
21	2.71
22	2.76
23	2.81
24	2.86
25	2.91
26	2.96
27	3.00
28	3.05
29	3.09
30	3.14

Source: CCITT. Recommendation E. 521, "Calculation of the Number of Circuits in a Group Carrying Overflow Traffic," Fourth Plenary Assembly, *White Book*, 1969. Reproduced with special authorization of the International Telecommunication Union (ITU).

maximum peakedness ratio of 2.05 given in table 5–2 for $c = 10$. Then table 5–3 shows that 7 circuits are needed to carry the *overflow* traffic of average value 0.166 erlangs at a blocking probability of 0.01. (In contrast, table 5–1 shows that only 3 circuits are needed to carry *random* traffic of average value 0.166 erlangs with this blocking probability.) In this example, the overall blocking probability of the combination of 10 circuits carrying the direct (random) traffic and 7 circuits carrying the overflow traffic is approximately $0.166 \times 0.01/5.53 = 0.0003$.

In practical networks with alternate routing, the overflows from several trunk groups that are each offered random traffic are in turn offered to one alternate route. For example, in figure 5–2, both the 1–2 and 1–4 (direct) trunk groups may overflow to the 1–3 group. A general picture of this kind of

Circuit-Switched Networks

network is shown in figure 5–4. To apply the CCITT method to find the size of the alternate-route trunk group, we first find the maximum peakedness ratio for each of the direct trunk groups from table 5–2; let us denote this ratio for direct trunk group k in figure 5–4 by r_k. Next we calculate the average overflow traffic for each direct trunk group, $b_k = a_k B(c_k, a_k)$. The *weighted mean peakedness ratio* is

$$r = \frac{\sum_1^n r_k b_k}{\sum_1^n b_k}$$

This value r is then used in table 5–3 to size the alternate-route trunk group; as noted earlier, the sizing in table 5–3 is for a blocking probability of 0.01 for the total overflow traffic. An example of the calculation of r is given in table 5–4, for $n = 3$ direct groups. From table 5–3, we find that 11 circuits are needed in the alternate-route group to carry the total overflow traffic of average value 1.421 erlangs with a blocking probability of 0.01.

In the event that the "alternate route" in figure 5–4 itself has some random offered traffic, this can be combined with the overflow traffic amounts in the method exemplified in table 5–4 simply by adding this random amount to the overflow total and taking its peakedness ratio as 1. For example, if, in table 5–4, there were 3.0 erlangs of random traffic offered directly to the alternate-route group, the total average offered traffic for this

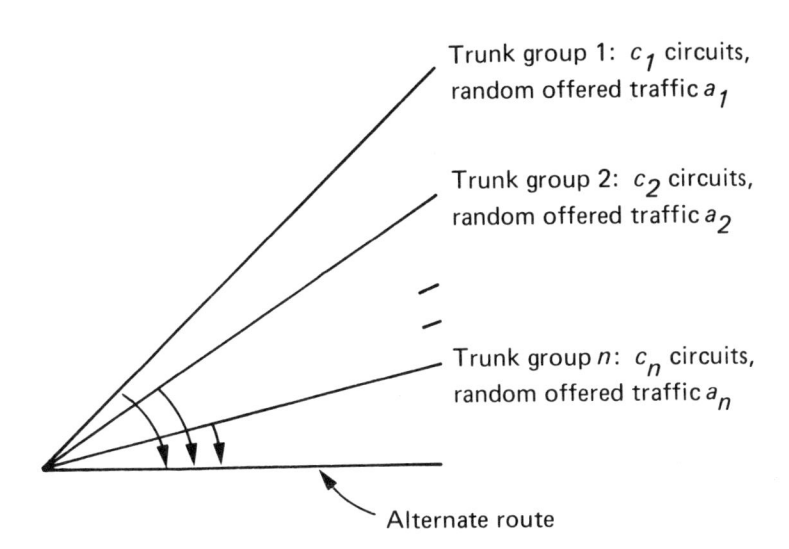

Figure 5–4. Trunk Groups 1 through n Overflowing to One Alternate Route

Table 5-3
Number of Circuits for a Blocking Probability $p = .01$ for Overflow Traffic
(Overflow Traffic in Erlangs)

Required Number of Circuits	\multicolumn{11}{c}{Maximum Peakedness Ratio or Weighted Mean Peakedness Ratio}
	1.0	1.1	1.2	1.3	1.4	1.5	1.6	1.7	1.8	1.9	2.0
1	0.01	0.0	0.0	0.0	0.0	0.0	0.0	0.0	0.0	0.0	0.0
2	0.15	0.03	0.0	0.0	0.0	0.0	0.0	0.0	0.0	0.0	0.0
3	0.46	0.32	0.19	0.04	0.0	0.0	0.0	0.0	0.0	0.0	0.0
4	0.87	0.72	0.57	0.42	0.28	0.11	0.0	0.0	0.0	0.0	0.0
5	1.36	1.20	1.03	0.87	0.71	0.55	0.39	0.22	0.0	0.0	0.0
6	1.91	1.73	1.55	1.38	1.20	1.04	0.87	0.70	0.53	0.35	0.15
7	2.50	2.30	2.11	1.92	1.74	1.56	1.38	1.21	1.03	0.86	0.68
8	3.13	2.91	2.71	2.50	2.31	2.12	1.93	1.75	1.57	1.38	1.21
9	3.78	3.55	3.33	3.12	2.91	2.71	2.51	2.32	2.12	1.94	1.75
10	4.46	4.22	3.98	3.76	3.54	3.32	3.11	2.91	2.71	2.51	2.32
11	5.16	4.90	4.65	4.42	4.18	3.96	3.74	3.53	3.32	3.11	2.91
12	5.88	5.60	5.34	5.09	4.85	4.61	4.38	4.16	3.94	3.73	3.52
13	6.61	6.32	6.05	5.78	5.53	5.28	5.05	4.81	4.59	4.36	4.15
14	7.35	7.05	6.77	6.49	6.23	5.97	5.72	5.48	5.24	5.01	4.79
15	8.11	7.80	7.50	7.21	6.94	6.67	6.41	6.16	5.92	5.68	5.44
16	8.88	8.55	8.24	7.95	7.66	7.38	7.12	6.85	6.60	6.35	6.11
17	9.65	9.32	9.00	8.69	8.39	8.11	7.83	7.56	7.30	7.04	6.79
18	10.44	10.09	9.76	9.44	9.13	8.84	8.55	8.27	8.00	7.24	7.48
19	11.23	10.87	10.53	10.20	9.88	9.58	9.28	9.00	8.72	8.45	8.18
20	12.03	11.66	11.31	10.97	10.64	10.33	10.02	9.73	9.44	9.16	8.89
21	12.84	12.46	12.09	11.75	11.41	11.09	10.77	10.47	10.18	9.89	9.61
22	13.65	13.26	12.89	12.53	12.18	11.85	11.53	11.22	10.92	10.62	10.34
23	14.47	14.07	13.68	13.32	12.96	12.62	12.29	11.97	11.66	11.36	11.07
24	15.29	14.88	14.49	14.11	13.75	13.40	13.06	12.73	12.42	12.11	11.81
25	16.12	15.70	15.30	14.91	14.54	14.18	13.84	13.50	13.18	12.86	12.56
26	16.96	16.53	16.11	15.72	15.34	14.97	14.62	14.28	13.94	13.62	13.31
27	17.80	17.35	16.93	16.53	16.14	15.77	15.41	15.06	14.72	14.39	14.07
28	18.64	18.19	17.76	17.34	16.95	16.57	16.20	15.84	15.49	15.16	14.83
29	19.49	19.02	18.58	18.16	17.76	17.37	16.99	16.63	16.28	15.93	15.60

Circuit-Switched Networks

Required number of Circuits	2.1	2.2	2.3	2.4	2.5	2.6	2.8	3.0	3.2	3.4	3.6	3.8	4.0
30	20.34	19.87	19.42	18.99	18.57	18.18	17.79	17.42	17.06	16.71	16.37		
31	21.19	20.71	20.25	19.81	19.39	18.99	18.60	18.22	17.85	17.50	17.15		
32	22.05	21.56	21.09	20.65	20.22	19.80	19.41	19.02	18.65	18.29	17.93		
33	22.91	22.41	21.93	21.48	21.04	20.63	20.22	19.83	19.45	19.08	18.72		
34	23.77	23.27	22.78	22.32	21.88	21.45	21.04	20.64	20.25	19.88	19.51		
35	24.64	24.12	23.63	23.16	22.71	22.28	21.86	21.45	21.06	20.68	20.31		
36	25.51	24.98	24.48	24.01	23.55	23.11	22.68	22.27	21.87	21.48	21.10		
37	26.38	25.85	25.34	24.85	24.39	23.94	23.51	23.09	22.68	22.29	21.91		
38	27.25	26.71	26.20	25.70	25.23	24.78	24.34	23.91	23.50	23.10	22.71		
39	28.13	27.58	27.06	26.56	26.08	25.61	25.17	24.74	24.32	23.91	23.52		
40	29.01	28.45	27.92	27.41	26.92	26.46	26.00	25.57	25.14	24.73	24.33		
41	29.89	29.32	28.79	28.27	27.78	27.30	26.84	26.40	25.97	25.55	25.15		
42	30.77	30.20	29.65	29.13	28.63	28.15	27.68	27.23	26.80	26.37	25.96		
43	31.66	31.08	30.52	30.00	29.49	29.00	28.53	28.07	27.63	27.20	26.78		
44	32.54	31.96	31.40	30.86	30.35	29.85	29.37	28.91	28.46	28.03	27.61		
45	33.43	32.84	32.27	31.73	31.21	30.70	30.22	29.75	29.30	28.86	28.43		
46	34.32	33.72	33.14	32.60	32.07	31.56	31.07	30.60	30.14	29.69	29.26		
47	35.21	34.61	34.02	33.47	32.93	32.42	31.92	31.44	30.98	30.53	30.09		
48	36.11	35.49	34.90	34.34	33.80	33.28	32.78	32.29	31.82	31.37	30.92		
49	37.00	36.38	35.78	35.21	34.67	34.14	33.63	33.14	32.29	34.67	32.21		31.26
50	37.90	37.27	36.67	36.09	35.54	35.01	34.49	34.00	33.52	33.05	34.59		

Maximum Peakedness Ratio or Weighted Mean Peakedness Ratio

Required number of Circuits	2.1	2.2	2.3	2.4	2.5	2.6	2.8	3.0	3.2	3.4	3.6	3.8	4.0
1	0.0	0.0	0.0	0.0	0.0	0.0	0.0	0.0	0.0	0.0	0.0	0.0	0.0
2	0.0	0.0	0.0	0.0	0.0	0.0	0.0	0.0	0.0	0.0	0.0	0.0	0.0
3	0.0	0.0	0.0	0.0	0.0	0.0	0.0	0.0	0.0	0.0	0.0	0.0	0.0
4	0.0	0.0	0.0	0.0	0.0	0.0	0.0	0.0	0.0	0.0	0.0	0.0	0.0
5	0.0	0.0	0.0	0.0	0.0	0.0	0.0	0.0	0.0	0.0	0.0	0.0	0.0
6	0.0	0.0	0.0	0.0	0.0	0.0	0.0	0.0	0.0	0.0	0.0	0.0	0.0
7	0.50	0.0	0.0	0.0	0.0	0.0	0.0	0.0	0.0	0.0	0.0	0.0	0.0
8	1.03	0.31	0.0	0.0	0.24	0.0	0.0	0.0	0.0	0.0	0.0	0.0	0.0
9	1.57	0.84	0.66	0.46	0.82	0.63	0.0	0.0	0.0	0.0	0.0	0.0	0.0
10	2.13	1.38	1.20	1.01	1.38	1.19	0.80	0.35	0.0	0.0	0.0	0.0	0.0
11	2.71	1.94	1.75	1.56	1.94	1.75	1.36	0.97	0.54	0.0	0.0	0.0	0.0
12	3.31	2.52	2.32	2.13	2.52	2.32	1.93	1.55	1.15	0.73	0.0	0.0	0.0
13	3.93	3.11	2.91	2.71	3.11	2.91	2.51	2.12	1.73	1.33	0.91	0.39	0.0
14	4.57	3.72	3.52	3.31	3.72	3.51	3.11	2.71	2.31	1.91	1.51	1.09	0.61

Table 5-3 *(continued)*

15	5.21	4.99	4.77	4.55	4.34	4.13	3.71	3.30	2.90	2.50	2.10	1.70	1.27
16	5.88	5.64	5.42	5.19	4.97	4.75	4.33	3.91	3.50	3.10	2.69	2.29	1.88
17	6.55	6.31	6.07	5.84	5.62	5.39	4.96	4.53	4.11	3.70	3.29	2.89	2.48
18	7.23	6.99	6.74	6.51	6.27	6.05	5.60	5.16	4.73	4.31	3.90	3.49	3.08
19	7.93	7.67	7.42	7.18	6.94	6.71	6.25	5.80	5.36	4.93	4.51	4.09	3.68
20	8.63	8.37	8.11	7.86	7.62	7.38	6.91	6.45	6.00	5.56	5.13	4.71	4.29
21	9.34	9.07	8.81	8.55	8.30	8.06	7.57	7.11	6.65	6.20	5.76	5.33	4.91
22	10.06	9.78	9.52	9.25	9.00	8.74	8.25	7.77	7.31	6.85	6.40	5.97	5.53
23	10.78	10.50	10.23	9.96	9.70	9.44	8.93	8.45	7.97	7.51	7.05	6.61	6.17
24	11.52	11.23	10.95	10.68	10.41	10.14	9.63	9.13	8.64	8.17	7.71	7.25	6.81
25	12.26	11.96	11.68	11.40	11.12	10.85	10.33	9.82	9.32	8.84	8.37	7.91	7.45
26	13.00	12.70	12.41	12.12	11.84	11.57	11.03	10.51	10.9	9.51	9.04	8.57	8.11
27	13.75	13.45	13.15	12.86	12.57	12.29	11.74	11.21	10.70	10.20	9.71	9.23	8.76
28	14.51	14.20	13.89	13.60	13.30	13.02	12.46	11.92	11.40	10.89	10.39	9.91	9.43
29	15.27	14.96	14.64	14.34	14.04	13.75	13.18	12.63	12.10	11.58	11.08	10.59	10.10
30	16.04	15.72	15.40	15.09	14.79	14.49	13.91	13.35	12.81	12.28	11.77	11.27	10.78
31	16.81	16.48	16.16	15.85	15.54	15.23	14.65	14.08	13.53	12.99	12.47	11.96	11.46
32	17.59	17.25	16.93	16.61	16.29	15.98	15.38	14.81	14.25	13.70	13.17	12.65	12.15
33	18.37	18.03	17.70	17.37	17.05	16.74	16.13	15.54	14.97	14.42	13.88	13.36	12.84
34	19.16	18.81	18.47	18.14	17.81	17.50	16.88	16.28	15.70	15.14	14.59	14.06	13.54
35	19.94	19.59	19.25	18.91	18.58	18.26	17.63	17.02	16.43	15.87	15.31	14.77	14.24
36	20.74	20.38	20.03	19.69	19.35	19.02	18.39	17.77	17.17	16.59	16.03	15.49	14.95
37	21.53	21.17	20.81	20.47	20.13	19.79	19.15	18.52	17.92	17.33	16.76	16.20	15.66
38	22.33	21.96	21.60	21.25	20.90	20.57	19.91	19.28	18.66	18.07	17.49	16.93	16.38
39	23.14	22.76	22.40	22.04	21.69	21.34	20.68	20.04	19.41	18.81	18.22	17.65	17.10
40	23.94	23.56	23.19	22.83	22.47	22.12	21.45	20.80	20.17	19.56	18.96	18.38	17.82
41	24.75	24.36	23.99	23.62	23.26	22.91	22.23	21.56	20.92	20.31	19.70	19.12	18.55
42	25.56	25.17	24.79	24.42	24.05	23.70	23.00	22.33	21.69	21.06	20.45	19.86	19.28
43	26.38	25.98	25.60	25.22	24.85	24.49	23.78	23.11	22.45	21.82	21.20	20.60	20.01
44	27.19	26.79	26.40	26.02	25.65	25.28	24.57	23.88	23.22	22.57	21.95	21.34	20.75
45	28.01	27.61	27.21	26.82	26.45	26.07	25.36	24.66	23.99	23.34	22.71	22.09	21.49
46	28.84	28.43	28.03	27.63	27.25	26.87	26.15	25.44	24.76	24.10	23.47	22.84	22.24
47	29.66	29.25	28.84	28.44	28.06	27.68	26.94	26.23	25.54	24.87	24.23	23.60	22.98
48	30.49	30.07	29.66	29.26	28.86	28.48	27.23	27.01	26.32	25.64	24.99	24.36	23.74
49	31.32	30.89	30.48	30.07	29.68	29.29	28.53	27.80	27.10	26.42	25.76	25.11	24.49
50	32.15	31.72	31.30	30.89	30.49	30.10	29.33	28.60	27.89	27.20	26.53	25.88	25.25

Source: CCITT. Recommendation E. 521, "Calculation of the Number of Circuits Carrying Overflow Traffic," Fourth Plenary Assembly, *White Book*, 1969. Reproduced with special authorization of the International Telecommunication Union (ITU).

Table 5-4
Example of Calculation of Weighted Mean Peakedness Ratio

k	a_k = Random Offered Traffic, Erlangs	c_k = Number of Circuits in Direct Group	Blocking Probability, $B(c_k, a_k)$[a]	Average Overflow Traffic, $b_k = a_k B(c_k, a_k)$	r_k = Maximum Peakedness Ratio[b]	$r_k b_k$
1	6.22	10	0.05	0.311	2.05	0.638
2	2.22	5	0.05	0.111	1.64	0.182
3	20.0	25	0.05	1.000	2.91	2.91
				1.421		3.73

[a]Table 5-1.
[b]Table 5-2.
r = weighted mean peakedness ratio = 3.73/1.421 = 2.62

group would be $1.421 + 3.0 = 4.421$, and the weighted mean peakedness ratio would be $(3.73 + 3.0)/4.421 = 1.52$. In general, as the direct traffic that is mixed with overflows increases, the weighted mean peakedness ratio decreases; when the direct traffic is dominant, it is not necessary to consider the peakedness of the combined traffic at all.

This simplified CCITT method is suited only for single-stage alternate routing, in which each direct traffic amount has only one alternate route. More recently, the CCITT has issued revised tables to replace table 5–3 that take into account both the degree of day-to-day variation in the busy-hour traffic as well as the finite length of the traffic observation period [10]. The effect of these revisions is a slight increase in the traffic capacity of a given number of circuits. In multistage networks in which the overflow from the first alternate route in turn overflows to a second alternate route, and so on, more complex methods of calculation based on the application of equation 5.9 are needed [7].

5–5 Blocking Probabilities in Networks

In this section, we deal with the problem of finding the blocking probabilities in networks with alternate routing. Whereas the CCITT method described in section 5–4 can be applied to a few trunk groups with alternate routing, the method of this section is more suitable for complex networks.

Consider the network of figure 5–5, in which there is offered traffic between each pair of nodes. Also there is a simple alternate-routing scheme, shown in the routing table in the figure, in which any traffic that is blocked on the direct link between two nodes is offered to the obvious second-choice route. We wish to find the *node-pair blocking probabilities* P_{ij}, that is, the proportion of i to j calls that is blocked for each node pair i,j. We will get at these values by first finding the *link-blocking probabilities*

$$p_{ij} = \frac{\text{traffic blocked on link } (i,j)}{\text{traffic offered to link } (i,j)}$$

We notice, first of all, that each link will carry some direct traffic and some overflow traffic. For example, link $(1, 3)$ carries part of a_{13} as direct traffic, as well as overflow traffic resulting from the blocking of both 1 to 2 calls (a_{12}) and 2 to 3 calls (a_{23}). Thus to find p_{13}, we should, as explained in section 5–4, take into account the peakedness of the two overflow traffic amounts. However, we assume here that overflow traffic is indistinguishable from random or direct traffic, or in other words, that the peakedness ratio of any overflow traffic amount is 1. (We discuss later methods that do not rely

Circuit-Switched Networks 75

Figure 5–5. Network with Alternate Routing

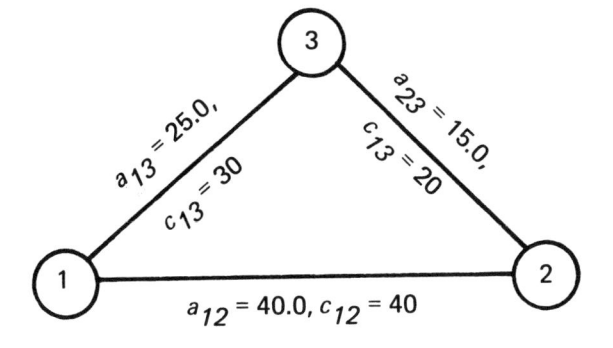

a_{ij} = busy-hour offered traffic from node i to node j, erlangs

c_{ij} = capacity (number of circuits) in link (i,j)

Routing Table

	First-choice route	Second-choice (alternate) route
a_{12}	(1,2)	(1,3), (3,2)
a_{13}	(1,3)	(1,2), (2,3)
a_{23}	(2,3)	(2,1), (1,3)

Network shown with: $a_{13} = 25.0$, $c_{13} = 30$; $a_{23} = 15.0$, $c_{23} = 20$; $a_{12} = 40.0$, $c_{12} = 40$.

on this assumption.) This assumption greatly simplifies the calculations since we now have

$$p_{ij} = B(c_{ij}, A_{ij}) \qquad (5.10)$$

Here A_{ij} is the total amount of traffic offered to link (i,j).

We can illustrate a method for finding the p_{ij} by referring again to figure 5–5. We make the further assumption that the link-blocking probabilities $p_{1\,2}$, $p_{1\,3}$, and $p_{2\,3}$ are statistically independent. (This assumption is not necessarily met in practice since blocking in one part of the network may be correlated with blocking in a nearby part. See also the later discussion in this section.) Also we denote by q_{ij} the *link-completion probability*, that is, $1 - p_{ij}$. If we take node pair 1, 2 as an example, we may now state that the probability $Q_{1\,2}$

of successfully completing a call from 1 to 2 (or vice versa), that is, the *node-pair completion probability*, is

$$1 - P_{12} = Q_{12} = q_{12} + q_{13}q_{23}(1 - q_{12}) \quad (5.11a)$$

In words, this states that a call is completed from 1 to 2 if link (1, 2) is not blocked, or if link (1, 2) is blocked and both links (1, 3) and (2, 3) are not blocked. We may further infer from equation 5.11a that of the total offered traffic a_{12} between 1 and 2, an amount $a_{12}Q_{12}$ is carried on the two routes available to it and that this is divided such that $a_{12}q_{12}$ is carried on link (1, 2) and $a_{12}q_{13}q_{23}(1 - q_{12})$ is carried on the alternate route, that is, on links (1, 3) and (2, 3). The corresponding equations for the other node-pair completion probabilities are

$$1 - P_{13} = Q_{13} = q_{13} + q_{12}q_{23}(1 - q_{13}) \quad (5.11b)$$

and

$$1 - P_{23} = Q_{23} = q_{23} + q_{12}q_{13}(1 - q_{23}) \quad (5.11c)$$

Thus the total carried traffic on link (i, j), which we will call A'_{ij}, may be expressed as in the following equations:

$$A'_{12} = a_{12}q_{12} + a_{13}q_{12}q_{23}(1 - q_{13}) + a_{23}q_{12}q_{13}(1 - q_{23}) \quad (5.12a)$$

$$A'_{13} = a_{13}q_{13} + a_{12}q_{13}q_{23}(1 - q_{12}) + a_{23}q_{12}q_{13}(1 - q_{23}) \quad (5.12b)$$

$$A'_{23} = a_{23}q_{23} + a_{12}q_{13}q_{23}(1 - q_{12}) + a_{13}q_{12}q_{23}(1 - q_{13}) \quad (5.12c)$$

Now, by applying equations 5.3 and 5.10, we have, for example,

$$q_{12} = 1 - B(c_{12}, A_{12}) = 1 - B\left(c_{12}, \frac{A'_{12}}{q_{12}}\right) \quad (5.13a)$$

The corresponding equations for links (1, 3) and (2, 3) are

$$q_{13} = 1 - B\left(c_{13}, \frac{A'_{13}}{q_{13}}\right) \quad (5.13b)$$

Circuit-Switched Networks

$$q_{23} = 1 - B\left(c_{23}, \frac{A'_{23}}{q_{23}}\right) \tag{5.13c}$$

We may now see that the problem of finding the link-blocking probabilities p_{ij} is equivalent to finding a set of q_{ij} values that is consistent with both equations 5.12 and 5.13.

A recursive procedure to find the q_{ij} is as follows. We assume an arbitrary set of initial values $q_{12}^{(0)}$, $q_{13}^{(0)}$, and $q_{23}^{(0)}$. Next we substitute these in equation 5.12 and calculate initial values $A'_{12}{}^{(0)}$, $A'_{13}{}^{(0)}$, and $A'_{23}{}^{(0)}$. Next we substitute these $A'_{ij}{}^{(0)}$ values in equation 5.13 and find new values $q_{12}^{(1)}$, $q_{13}^{(1)}$, and $q_{23}^{(1)}$. If $q_{12}^{(1)}$ is equal to $q_{12}^{(0)}$, or close enough to it to meet a suitable stopping criterion, and this is also true of $q_{13}^{(1)}$ and $q_{23}^{(1)}$ in relation to $q_{13}^{(0)}$ and $q_{23}^{(0)}$, then the procedure ends. Otherwise, we use the $q_{ij}^{(1)}$ in equation 5.12 and continue, finding new values $A'_{ij}{}^{(1)}$, and so forth. This numerical procedure has been found to converge in practical networks in a reasonable number of steps; that is, after k steps there is little or no change from $q_{ij}^{(k)}$ to $q_{ij}^{(k+1)}$. We use the terminating probabilities $q_{ij}^{(k)}$ in equation 5.11 to find the node-pair completion probabilities Q_{ij} for each parcel of offered node-to-node traffic a_{12}, a_{13}, and a_{23}.

As an example, suppose that in figure 5–5, we use the arbitrary initial values $q_{12}^{(0)} = q_{13}^{(0)} = q_{23}^{(0)} = 1$, that is, we assume that all three link-blocking probabilities are zero. When we substitute these values in equation 5.12, we find $A'_{12}{}^{(0)} = a_{12} = 40.0$, $A'_{13}{}^{(0)} = a_{13} = 25.0$, and $A'_{23}{}^{(0)} = a_{23} = 15.0$, as we would expect with no link blocking. When we substitute these $A'_{ij}{}^{(0)}$ values in equation 5.13, we have

$$q_{12}^{(1)} = 1 - B(40, 40.0) = 0.8838$$

$$q_{13}^{(1)} = 1 - B(30, 25.0) = 0.9474$$

$$q_{23}^{(1)} = 1 - B(20, 15.0) = 0.9544$$

As the next step, we substitute the $q_{ij}^{(1)}$ in equation 5.12, and we find $A'_{12}{}^{(1)} = 37.0$, $A'_{13}{}^{(1)} = 28.5$, and $A'_{23}{}^{(1)} = 19.6$. Continuing, we find that the q_{ii} values stabilize after 16 steps: $q_{12}^{(16)} = 0.7853$, $q_{13}^{(16)} = 0.7898$, and $q_{23}^{(16)} = 0.6997$. For the node-pair blocking probabilities, we have, after this many iterations, $1 - Q_{12} = 0.096$, $1 - Q_{13} = 0.096$, and $1 - Q_{23} = 0.115$.

Generalization

For the simple network of figure 5–5, it was possible to derive the node-pair completion probabilitites, Q_{ij}, as in equation 5.11, and the link carried traffic values, as in equation 5.12, by inspection. To apply our heuristic method for finding the link-completion probabilities q_{ij} to more complex networks with alternate routing, we proceed as follows.

We will concentrate on a particular node pair i, j. The *routing table* for this node pair is an ordered list of routes $R_1, R_2, \ldots, R_n$. Our procedure applies to a deterministic routing scheme, in which each call from i to j is offered first to R_1, and if this route is busy, that is, if there is not at least one idle circuit in each of its links, the call is next offered to R_2, and so forth. The call is blocked if and only if all routes R_1 through R_n are busy.

Each route R_m from i to j is described by the list of links it contains. For example, in figure 5-5, referring to node pair 1, 3, R_1 consists of link (1, 3) and R_2 of links (1, 2) and (2, 3). We will denote by x_m the product of the link-completion probabilities of the links in R_m. Continuing the same example from figure 5-5, $x_2 = q_{12}q_{23}$. Following Segal, the general node-pair completion probability [11] is

$$Q_{ij} = 1 - (1 - x_1) * (1 - x_2) * \cdots *(1 - x_n) \qquad (5.14)$$

Here the asterisks mean that the following operations must be carried out to evaluate Q_{ij} numerically. First, the x_m are substituted in equation 5.14 in symbolic form, that is, in terms of the q_{rs}. Next the right-hand side of equation 5.14 is expanded as if the * denoted ordinary multiplication, resulting, in general, in a polynomial in the q_{rs}. Next this polynomial is reduced according to the rule that no term q_{rs} may appear to a power higher than the first. Thus the rule is

$$q_{rs}^a q_{tu}^b = q_{rs} \text{ if } (r, s) \text{ is the same link as } (t, u); \qquad (5.15)$$
$$a \text{ and } b \text{ are exponents} \geq 1$$

We apply the rule of equation 5.15 here for the same reason as explained in section 4-5 in connection with the calculation of terminal reliability, that is, to obtain the correct value of probability Q_{ij} when a link is common to two or more routes. Finally, after the rule of equation 5.15 has been applied, we substitute numerical values of the q_{rs} to calculate Q_{ij}.

To illustrate the calculation of Q_{ij}, consider figure 5-6, which shows a fragment of a network and two routes between nodes a and d. If these are the only routes of interest we have

$$\begin{aligned} Q_{ad} &= 1 - (1 - x_1)*(1 - x_2) \\ &= 1 - (1 - q_{ab}q_{bd})*(1 - q_{ab}q_{bc}q_{cd}) \\ &= q_{ab}q_{bc}q_{cd} + q_{ab}q_{bd} - q_{ab}q_{bd}q_{bc}q_{cd} \end{aligned}$$

Circuit-Switched Networks

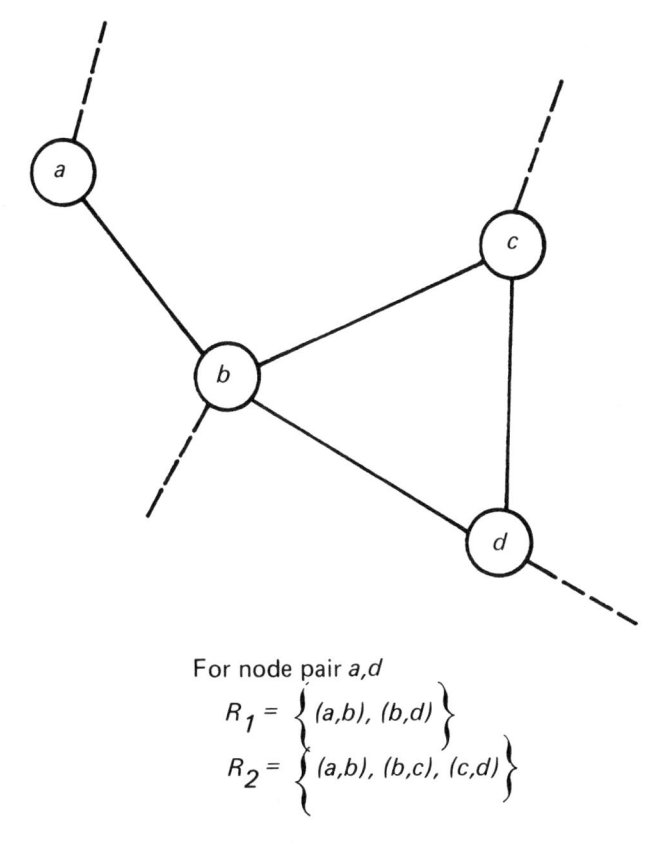

For node pair a,d
$$R_1 = \{(a,b), (b,d)\}$$
$$R_2 = \{(a,b), (b,c), (c,d)\}$$

Figure 5-6. Fragment of Network

By substituting numerical values of the q_{rs} in the right-hand side of the final line above, we would have the value of Q_{ad}.

To continue the development of a general procedure, we note that, just as equation 5.14 gives the node-pair completion probability for nodes i and j when all n alternate routes are considered, the truncated expression

$$Q_{ij,m} = 1 - (1 - x_1) * (1 - x_2) * \cdots * (1 - x_m) \qquad m \leq n \qquad (5.16)$$

is the probability that a call between i and j is completed on any one of the first m alternate routes. Thus the proportion of the offered traffic, a_{ij}, between i and j that is carried on the mth alternate route is

$$h_m = a_{ij}(Q_{ij,m} - Q_{ij,m-1}) \qquad m = 1, 2, \ldots, n \qquad (5.17)$$

in which $Q_{ij,0}$ is understood to be zero, and, of course, $Q_{ij,n}$ is identical to Q_{ij} in equation 5.14.

Let us now enlarge our view from the particular node pair i, j to the entire network, or more exactly, to the set of all node pairs for which there is offered traffic. For each node pair, say, r, s we have the routing table $R_1^{rs}, R_2^{rs}, \ldots, R_n^{rs}$; in practical networks, the total number of alternate routes n may vary with the node pair. From the set of routing tables, we find the values of Q_{rs} as in equation 5.14, the $Q_{rs, m}$ ($m \leq n$), as in equation 5.16, and the h_m^{rs}, that is, the proportions of the offered traffic a_{rs} carried on the mth alternate route for this node pair, as in equation 5.17. We may now find the total carried traffic on any link in the network, say, link (u, v), as

$$A'_{uv} = \sum_{r,s} \sum_{t=1}^{n} I_{rst} h_t^{rs} \tag{5.18}$$

Here h_t^{rs} is the proportion of the offered traffic a_{rs} for node pair r, s that is carried on the tth alternate route for this node pair, and I_{rst} is an index whose value is 1 if this tth alternate route includes link (u, v) and zero if it does not. The second, or inner, summation sign in equation 5.18 means that, for any node pair, r, s all n of its alternate routes must be included. The first, or outer, summation sign means that all (active) node pairs r, s must be included.

We may incorporate these calculations into an iterative procedure in the following.

Network Blocking Probability Algorithm

The following are given: the busy-hour offered traffic a_{ij} in erlangs for each node pair i, j[d]; the routing table for each node pair for which $a_{ij} > 0$; and the capacity c_{uv} of each link (u, v). The algorithm, which is heuristic, finds values for the link-completion probabilities q_{uv}, where $q_{uv} = 1 -$ [link-blocking probability for link (u, v)], and of the node-pair completion probabilities, Q_{ij}, where $Q_{ij} = 1 -$ (proportion of offered traffic from i to j that is blocked), under the assumptions that overflow traffic is indistinguishable from direct traffic and that the link-blocking probabilities are independent.

Step 1 [Set initial values of q_{uv}]. Let k be an iteration index and set $k \leftarrow 0$, $q_{uv}^{(k)} \leftarrow 1$ for all links (u, v).

---

[d]If the offered traffic is given as two components, one from i to j and one from j to i, we assign a_{ij} a value that is the sum of the two. It is implicit here that the routing table for node pair i, j is the same as that for node pair j, i, the routes being traversed in the opposite order.

Circuit-Switched Networks

Step 2 [Calculate carried traffic $A'^{(k)}_{uv}$ on each link]. Calculate the current value of the traffic carried on link (u, v), viz., $A'^{(k)}_{uv}$, for each link (u, v). This requires, first, the calculation of the node-pair completion probability Q_{ij} for each active node pair, that is, each node pair i, j for which $a_{ij} > 0$, using equation 5.14; the calculation of the truncated probabilities $Q_{ij,m}$ using equation 5.16, and the carried traffic contributions h_m using equation 5.17; and finally the summation of the h_m for all routes that traverse link (u, v) using equation 5.18.

Step 3 [Calculate new values of the q_{uv}]. Set $k \leftarrow k + 1$ and find

$$q_{uv}^{(k)} = 1 - B\left(c_{uv}, \frac{A'^{(k-1)}_{uv}}{q_{uv}^{(k-1)}}\right)$$

for all links (u, v).

Step 4 [Apply stopping criterion]. If for all links (u, v), $q_{uv}^{(k)}$ is sufficiently close to $q_{uv}^{(k-1)}$, for example, if

$$\max_{\text{all links } (u, v)} \left| \frac{q_{uv}^{(k)}}{q_{uv}^{(k-1)}} - 1 \right| \leq \varepsilon$$

where ε is a small positive value, for example 0.001, stop. Otherwise, go to step 2.

The node-pair completion probabilities Q_{ij}, calculated in step 2, may be used to derive indexes of overall network performance such as the average value of Q_{ij} over all (active) node pairs, $\Sigma\, Q_{ij} a_{ij} / \Sigma\, a_{ij}$, or the minimum value of Q_{ij} (maximum node-pair blocking value) throughout the network.

The algorithm, particularly in step 2, generalizes the steps that we carried out earlier for the simple network of figure 5–5 by inspection. To illustrate this, we note, for example, that $Q_{1\,2}$ in this network can be found by equation 5.14 as

$$Q_{1\,2} = 1 - (1 - x_1) * (1 - x_2)$$

where

$$x_1 = q_{1\,2}$$

and

$$x_2 = q_{1\,3} q_{2\,3}$$

Thus

$$Q_{12} = 1 - (1 - q_{12})*(1 - q_{13}q_{23}) = q_{12} + q_{13}q_{23}(1 - q_{12})$$

just as in equation 5.11a. Also, $Q_{12,1}$, the proportion of the 1–2 traffic carried on the first-choice route, is by equation 5.16,

$$Q_{12,1} = 1 - (1 - x_1) = q_{12}$$

just as we observed in the explanation following equation 5.11a. Continuing, we note that equation 5.12a, for example, corresponds to equation 5.18, with $uv = 12$, $h_1^{12} = a_{12}q_{12}$, $h_2^{13} = a_{13}q_{12}q_{23}(1 - q_{13})$, and so on, and $I_{121} = I_{132} = 1$, and so on.

Critique of the Method

The author programmed the network blocking probability algorithm as an application of Segal's method [11] to the analysis of networks with as many as 150 nodes, 255 links, and 8 alternate routes per node pair, principally as an alternative to simulation.[e] Covo has reported the results of this work and its extensions to networks with priorities and preemption [12, 13]. The node-pair blocking probabilities found with the algorithm agree reasonably well with those found in simulations. Nevertheless, the algorithm rests upon two unsupportable assumptions: (a) that overflow traffic has the same peakedness rato as directly offered (Poisson) traffic; and (b) that the link-blocking probabilities are statistically independent. As to assumption (a), we know, as discussed in section 5–4, that overflow traffic is more peaked than direct traffic, that is, that the ratio of its variance to its mean is greater than 1. That the independence assumption (b) does not always conform to reality is affirmed by Holtzman, who also gives an initial approach to accounting for dependencies among link-blocking probabilities [14].

Several investigators give procedures that do not require assumption (a) about the similarity of overflow and direct traffic, although they retain the independence assumption (b). Katz estimates the mean and variance of the overflow traffic amounts separately [15], whereas Kuczura and Bajaj estimate the first three moments of the overflow traffic amounts separately [16]. Table 5–5 shows the node-pair blocking probabilities for the network of figure 5–5 found by three different methods: the algorithm of this section, the

---

[e]Apart from their theoretical interest, analytic methods tend to require less computer time to run than simulations. In our further discussion in this section, we regard the simulation results as a standard of comparison because, properly run, a simulation can yield more precise results.

Table 5-5
Node-pair Blocking Probabilities for Network of Figure 5-5

	Method		
Node Pair	Network Blocking Probability Algorithm	Kuczura and Bajaj[a] Three-Moment	Kuczura and Bajaj[a] Simulation
1, 2	.096	.086	.089
1, 3	.096	.085	.082
2, 3	.115	.10	.098

[a]A. Kuczura and D. Bajaj. "A Method of Moments for the Analysis of a Switched Communication Network's Performance." *IEEE Trans. Commun.* 25:185–193, 1977.

three-moment method of Kuczura and Bajaj, and a simulation reported by the same authors. Although the results of the three-moment method conform more closely to those of the simulation than do the results of the network blocking probability algorithm, the latter results are precise enough for many practical situations.

References

1. Syski, R. *Introduction to Congestion Theory in Telephone Systems.* Edinburgh: Oliver and Boyd, 1960.
2. Hayward, W.S., Jr. "The Reliability of Telephone Traffic Load Measurements by Switch Counts," *Bell Syst. Tech. J.* 31:357–377, 1952.
3. Brockmeyer, E. et al. *The Life and Works of A.K. Erlang.* Acta Polytechnica Scandinavica, Applied Mathematics and Computer Machinery Series, no. 6, 1960.
4. Jagerman, D.L. "Some Properties of the Erlang Loss Function," *Bell Syst. Tech. J.* 53:525–551, 1974.
5. Mina, R.R. *Introduction to Teletraffic Engineering.* Chicago: Telephony Publishing, 1974.
6. Frankel, T. *Tables for Traffic Management and Design.* Chicago, Lee's abc of the Telephone, 1976.
7. Wilkinson, R.I. "Theories for Toll Traffic Engineering in the U.S.A.," *Bell Syst. Tech. J.* 35:421–514, 1956.
8. Jewett, J.E. "Changes in Optimal Long Distance Network Policies," *Bus. Commns. Rev.*, March/April 1976.
9. CCITT. Recommendation E. 521, "Calculation of the Number of Circuits in a Group Carrying Overflow Traffic," Fourth Plenary Assembly, *White Book*, 1969.

10. CCITT. Recommendation E. 521, "Calculation of the Number of Circuits in a Group Carrying Overflow Traffic," Sixth Plenary Assembly, *Orange Book*, 1977.
11. Segal. "Traffic Engineering of Communications Networks with a General Class of Routing Schemes," Fourth International Teletraffic Conference, London, 1964.
12. Covo, A.A. "Sizing of Military Circuit-switched Communication Networks by Computer-aided Analytic Methods," International Conference on Communications, Seattle, 1973.
13. Covo, A.A. "Analysis of Circuit-switched Communication Networks with Multiple Priorities and Preemption." *GTE J. Res. Dev.* 1:27–38, 1974.
14. Holtzman, J.M. "Analysis of Dependence Effects in Telephone Trunking Networks," *Bell Syst. Tech. J.* 50:2647–2662, 1971.
15. Katz, S. "Statistical Performance Analysis of a Switched Communications Network," Fifth International Teletraffic Congress, 1967.
16. Kuczura, A., and Bajaj, D. "A Method of Moments for the Analysis of a Switched Communication Network's Performance," *IEEE Trans. Commun.* 25:185–193, 1977.

6 Message-Switched Networks

6–1 Introduction

In a message-switched network, the user sends a message to a switching node where the message is temporarily stored. The switching equipment at the node determines from the address in the message the identity of the node to which to forward the message, (hence the equivalent term *store-and-forward* switching). It then selects a link to this next node and, when this link becomes available, forwards the message. This sequence is repeated until the message arrives at the destination node, which delivers it to the addressed user. For example, in figure 6–1, user A, to send a message to user B, first sends it to node 1, at which it is stored. If we assume that the preferred route is $\{(1, 2), (2, 4)\}$, node 1 sends it to node 2 when link $(1, 2)$ becomes available, and the message is now stored at node 2, pending the availability of link $(2, 4)$.

Message switching can achieve better utilization of the links than circuit switching (see chapter 5) at the cost of a delay in the delivery of messages; a message-switched network is not suitable for the immediate exchange of messages and replies, as is a circuit-switched network. Referring to figure 6–1, B will receive A's message after a delay that may, in practical systems, vary from a fraction of a second to several hours, measured from the time the message is first stored, at node 1. This delay consists of the times for the node switches to store and retrieve the message and to select and seize outgoing links; the times needed to transmit the message along the links; and the times that the message waits in storage queues at the nodes for links to become available. The first two types of delays, for node processing and link transmission, can be made small relative to the maximum tolerable delay. Thus the user-to-user delay, which is the principal criterion of performance, depends largely on the queueing delays at the nodes. In this chapter, we deal mainly with methods for estimating the node-to-node delay in the *backbone* network (see figure 6–1); thus we will assume that delays on the user links are not significant.

Whereas in a circuit-switched network each link can handle from a few to many thousand simultaneous user-to-user connections, a link in a message-switched network can be used by just one message at a time. A model of a message-switched network developed by Kleinrock, based on the theory of the single-server queue, has been widely accepted. We will first look at the

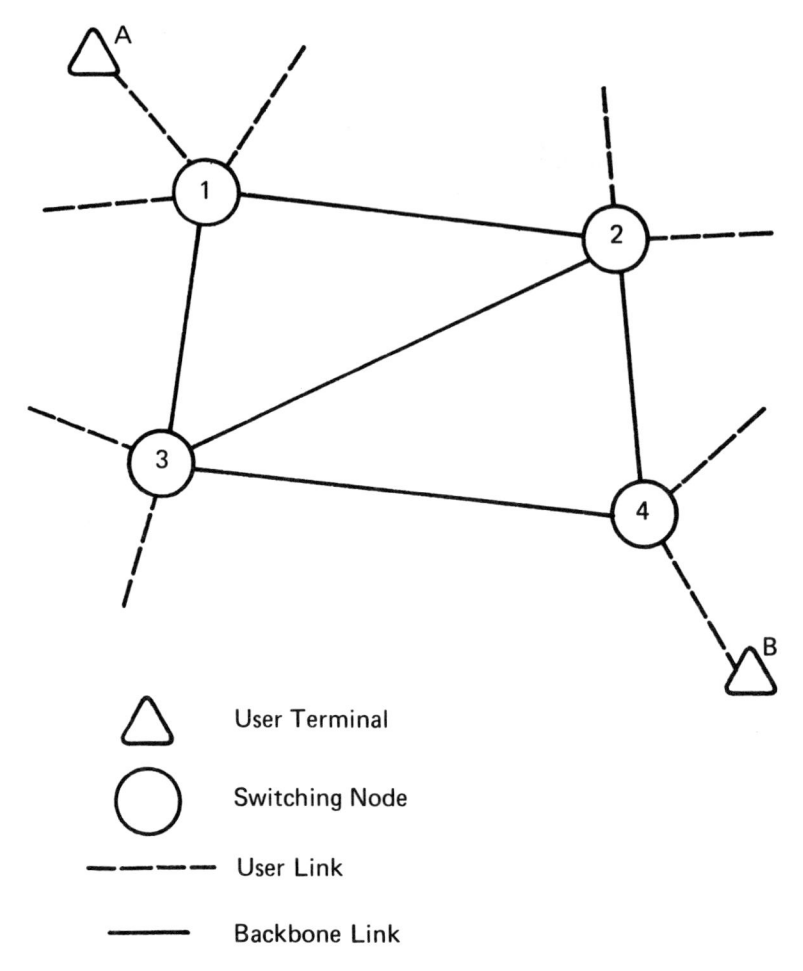

Figure 6-1. Message-switched Network

properties of the single-server queue and then at Kleinrock's model and its extensions.

6-2 The Single-server Queue

We will isolate one node, as shown in figure 6-2, and consider messages that are queued for transmission outward from this node on one unidirectional link, which is the "server." The actions at the node of storing and retrieving a message and of connecting to the link when it becomes available are idealized as taking no time, or at most an insignificant amount of time compared to the

Message-Switched Networks 87

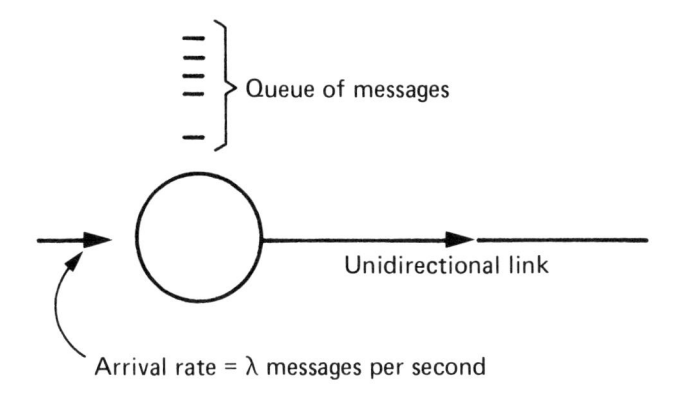

Figure 6-2. A Unidirectional Link as a Single-server Queue

average service time which is defined later in this section. The development here is analogous to that in section 5-2 for the full-availability trunk group, with the necessary changes for dealing with a delay system rather than a blocking system.

Assumptions

We assume that the arrivals of messages at the node have a *Poisson* distribution. Thus the probability that exactly j messages will arrive in an interval of t seconds is

$$p_j(t) = (\lambda t)^j \exp(-\lambda t)/j!, j = 0, 1, 2, \ldots \qquad (6.1)$$

In equation 6.1, λ is a positive constant with the following meaning: the intervals between successive message arrivals, that is, the *interarrival times*, have a negative exponential distribution with mean $1/\lambda$ seconds. Thus the probability that an interarrival time is less than or equal to t_1 seconds is $1 - \exp(-\lambda t_1)$, $t_1 \geq 0$. The average rate of message arrivals, that is, the *throughput*, is λ messages per second.

We assume that the lengths of the messages have a negative exponential distribution with mean $1/\mu$ bits, the unit of bits being chosen rather than seconds for our later convenience. Thus the probability that any message length is less than or equal to b bits is $1 - \exp(-\mu b)$, $b \geq 0$. We assume further that the capacity or transmission speed of the link is C bits per second. Thus the link transmission time, or, in queueing theory, the *service time*, for a message b bits in length is b/C seconds.

We assume that the stored messages are served, that is, transmitted, in order of their arrival; this is the first-in, first-out (FIFO) queue discipline. As a consequence of the assumption of Poisson arrivals, it is also necessary to assume that an arbitrarily large amount of storage is available in the nodes for messages in queue, although in practical systems the storage is limited. In any event, we will assume that there is sufficient storage at the nodes so that only rarely is it not possible to add a message to the queue.

We assume that the paramenters λ and μ that characterize the statistics of message arrivals and message lengths are constants, not dependent on the time. Furthermore, we assume that the system is in statistical equilibrium, which we can imagine has been attained by allowing messages to queue up and to be transmitted on the link for a time that is at least several times the average message service time, $1/C\mu$ seconds.

Average Time in the System

The delay for any message is the time it is held in storage at the node in figure 6–2 before the link becomes available; it may, of course, be zero. A fundamental result of queueing theory [1] is that the average delay for all messages is

$$\tau = \frac{\rho}{\mu C(1 - \rho)} \qquad (6.2)$$

where ρ is the *utilization factor*[a] for the link:

$$\rho = \frac{\lambda}{\mu C} \qquad (6.3)$$

Suppose, for example, that the arrival rate is $\lambda = 5$ messages per second, the average message length is $1/\mu = 100$ bits, and the link capacity is $C = 2,000$ bits per second. Then we have $\rho = (5 \times 100/2000) = 0.25$, and $\tau = 0.25/[(2000/100) \times (1 - 0.25)] = 0.017$ seconds.

In figure 6–2 each message spends some time in the storage queue at the node and additional time being transmitted on the link. The total of these two time intervals is the *time in the system*, and the average of this quantity for all messages, T, is just the sum of the average delay, τ, and the average service time, $1/\mu C$:

$$T = \tau + (1/\mu C) = 1/\mu C(1 - \rho) \qquad (6.4)$$

---

[a]The dimensionless utilization factor is exactly analogous to the offered traffic in erlangs; see chapter 5.

Message-Switched Networks

Thus in the previous example $T = 0.017 + (100/2,000) = 0.067$ seconds. We notice that as ρ approaches zero, τ approaches zero, and T is just the average service time, $1/\mu C$. For later convenience, we note that in view of the definition of ρ, equation 6.4 may be written

$$T = \frac{1}{(\mu C - \lambda)} \tag{6.5}$$

Since the expressions for both τ and T contain the term $1 - \rho$ in their denominators, we must have $\rho < 1$, or equivalently $\mu C > \lambda$, in any practical system. Otherwise the system is unstable in the sense that the average delay, and hence the average time in the system, increases beyond all bounds.

The queueing system of figure 6-2 with the assumptions described here is known as the M/M/1 system in a notation attributed to D.G. Kendall. Here the first letter M refers to the Poisson or random (Markov) nature of the distribution of message arrivals; a completely general distribution would be denoted by G. The second letter M refers to the negative exponential distribution of message lengths. Once again, if the message lengths had an arbitrary distribution, the letter G would appear. The final 1 is the number of servers, or in our example, the number of links. Thus, for example, the M/G/1 system has random message arrivals, a general distribution of message lengths, and one link. Whereas solutions have been found for this case and others that differ from M/M/1, we will not use them here [1, 2].

6-3 Kleinrock's Model

Kleinrock applied the results for the single-server queue, described in section 6-2, to a message-switched backbone network such as that shown in figure 6-1 [3].

The notation for the network problem is an extension of that which we have used in section 6-2, as follows. We denote by γ_{jk} the arrival rate of messages at origin node j with destination node k, in messages per second; these arrivals are assumed to occur at random, that is, with a Poisson distribution, as discussed in section 6-2. For example, in figure 6-1, γ_{14} equals the arrival rate at node 1 of messages from user A and other users connected to node 1, addressed to user B and other users connected to node 4. The quantity γ is the total arrival rate of messages into the backbone network from all users: $\gamma = \Sigma_{j,k} \gamma_{jk}, j \neq k$.[b]

---

[b]The model does not deal with messages between users connected to the same node. It is not necessary that $\gamma_{jk} = \gamma_{kj}, j \neq k$. Both j and k vary from 1 to the number of nodes in the network.

The model permits only *fixed routing*. That is, for any node pair j, k, there is just one route for all messages from j to k. Furthermore, each communications link between nodes m and n consists of two unidirectional links, one directed from m to n and called link (m, n), and the other directed from n to m and called (n, m). Thus the graph is *directed*, unlike most of the graphs that we deal with in this book. In figure 6–1, for example, we may choose for messages from node 1 to node 4 just one of the routes $\{(1, 2), (2, 4)\}$, $\{(1, 3), (3, 4)\}$, $\{(1, 2), (2, 3), (3, 4)\}$, or $\{(1, 3), (3, 2), (2, 3)\}$. The selected route need not be the shortest route. The route for j, k messages need not be the same as that found by traversing the k, j route in reverse link order.) Alternate routing is not permitted. With this limitation, we may define λ_i as the arrival rate of messages to the ith link; it is not difficult to see that λ_i is the sum of all γ_{jk} for which the (fixed) route from node j to node k includes link i. For example, if in figure 6–1 the total traffic on link $(2, 4)$ consists of all messages from node 1 to node 4 and all messages from node 3 to node 4, then the value of λ_i for this link is $\gamma_{14} + \gamma_{34}$. We will denote by λ the sum of the λ_i for all links: $\lambda = \Sigma_i \lambda_i$.

The quantity $1/\mu$, as before, denotes the average message length in bits; the message lengths are assumed, as before, to have a negative exponential distribution. In his analysis, Kleinrock found that there are severe mathematical difficulties if one requires that a message, once it is generated at a user terminal (see figure 6–1) retains the same length throughout its passage through the backbone network, as is actually the case. He therefore introduced a simplifying assumption that is contrary to reality but which he showed, by means of simulation tests, nevertheless allows the model to produce reliable results. This is the *independence assumption*: each time a message is received at a node of the backbone network, stored, and transmitted to the next node, a new length is chosen for it at random from a negative exponential distribution with average length $1/\mu$ bits. This is somewhat analogous to the assumption in section 5–5 that overflow traffic is indistinguishable from direct traffic in a circuit-switched network.

As a consequence of the independence assumption, of the random nature of message arrivals at the ndoes, and of a theorem in queueing theory [4], the queues at the nodes of messages waiting for links to become available can each be treated independently according to the single-server queue results of section 6–2. Thus the contribution to the average time in the system for messages that traverse link i is, from equation 6.5

$$T_i = \frac{1}{(\mu C_i - \lambda_i)} \tag{6.6}$$

where C_i is the capacity of link i, bits per second. Equation 6.6 shows that we must have $\mu C_i > \lambda_i$ for each link i in any practical system, just as

Message-Switched Networks

we have noted in connection with equation 6.5 for the isolated single-server queue.

To illustrate these calculations, in the three-node network of figure 6–3, the unidirectional links are shown explicitly and numbered arbitrarily from 1 to 4. We have $\lambda_1 = \gamma_{13} + \gamma_{12}$, $\lambda_2 = \gamma_{13} + \gamma_{23}$, $\lambda_3 = \gamma_{31} + \gamma_{32}$, and $\lambda_4 = \gamma_{21} + \gamma_{31}$. By applying equation 6.6 independently to the four links, we have as the contribution to the average time in the system for messages that traverse link 1,

$$T_1 = \frac{1}{\mu C_1 - (\gamma_{13} + \gamma_{12})}$$

for messages that traverse link 2

$$T_2 = \frac{1}{\mu C_2 - (\gamma_{13} + \gamma_{23})}$$

and so on. Thus the average time in the system for 1-to-2 messages is T_1, whereas that for 1-to-3 messages is $T_1 + T_2$.

An important aspect of Kleinrock's work is his definition of the *average time in the system for all messages*, which is the network-performance measure, as

$$T = \sum_i \frac{T_i \lambda_i}{\gamma} \tag{6.7}$$

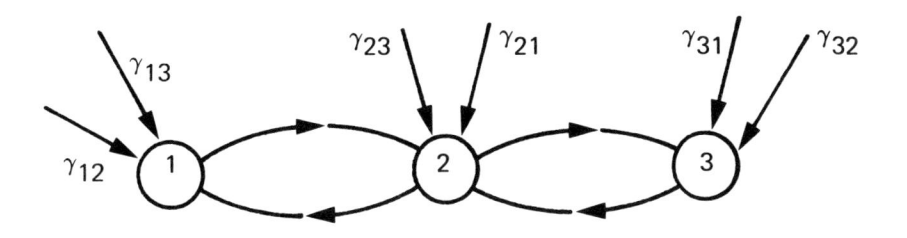

Link	Link Number
(1,2)	1
(2,3)	2
(3,2)	3
(2,1)	4

Figure 6–3. Three-node Network

in which the summation extends over all the unidirectional links in the network, and the T_i are found from equation 6.6. Kleinrock bases equation 6.7 on the definitions of the λ_i and T_i.

6–4 Network Optimization: The Capacity Assignment Problem

Much of the interest in Kleinrock's model arises from his further results on optimizing a message-switched network. That is, he showed how to choose the individual unidirectional link capacities C_i so as to minimize the average time in the system for all messages, T, as given by equation 6.7, under the constraint that the total link capacity $C = \Sigma_i C_i$ is fixed. This is called the *capacity assignment* (CA) problem. Of course, his solution for the optimum set of C_i applies under the assumptions of the model, viz., Poisson message arrivals at the nodes, fixed routing, and the independence assumption. The optimum capacity for the ith link is [3]

$$C_i = \left(\frac{\lambda_i}{\mu}\right) + C(1 - n\rho)\lambda_i^{1/2} \sum_j \lambda_j^{1/2} \qquad (6.8)$$

Here the quantity $n = \lambda/\gamma$ is the average number of links traversed for all messages. {This interpretation of n is not immediately obvious. It is established by Kleinrock using the definitions of the λ_i and γ [3].} Also, $\rho = \gamma/\mu C$. The contribution to the average time in the system for messages traversing link i, which is found by substituting from equation 6.8 into the definition, equation 6.6 is

$$T_i = \frac{\sum_j \lambda_j^{1/2}}{\mu C(1 - n\rho)\lambda_i^{1/2}} \qquad (6.9)$$

The average time in the system for all messages, which is found by substituting from equation 6.8 into equation 6.7, is

$$T = \frac{n\left[\sum_i (\lambda_i/\lambda)^{1/2}\right]^2}{\mu C(1 - n\rho)} \qquad (6.10)$$

For the set of capacities C_i given by equation 6.8 to constitute a feasible solution, each link must have enough capacity to sustain its average message rate, that is, we must have $C_i > \lambda_i/\mu$ for all links i, as we have noted following

Message-Switched Networks

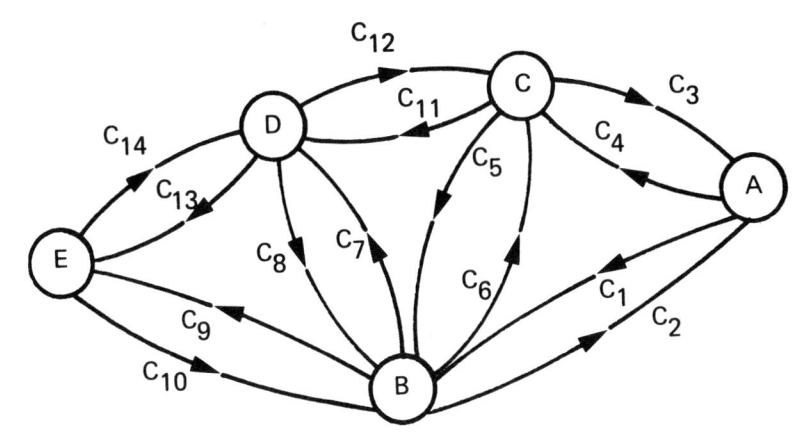

Figure 6–4. Network for Optimization Example

equation 6.6. Equation 6.9 further shows that we must have $\rho < 1/n$, or equivalently $\lambda < \mu C$, in order to keep the T_i, and thus T, within finite bounds. As ρ increases toward $1/n$, T has a simple pole; thus, with optimum link capacities all links saturate at exactly the same network load, γ. On the other hand, if the capacities are other than optimum, for example, if they are discrete capacities that do not match the optimum values, it is likely that there will be one link that is a bottleneck. That is, for some link m, the quantity $\rho_m = \lambda_m/\mu C_m$ will approach 1 sooner, as the network ρ increases, than for any other link. The behavior of T versus ρ will thus be dominated by the behavior of link m [2].

An Example

To illustrate Kleinrock's optimization method, we will refer to the network of figure 6–4, in which the unidirectional link capacities $C_1, C_2, \ldots, C_{14}$ are labeled explicitly. We assume a set of demands, that is, message arrival rates γ_{ij}, as shown in table 6–1.[c] It happens that this table is symmetric, that is, that $\gamma_{ij} = \gamma_{ji}$. We will also assume that the routing table is symmetric, for example, that the route for messages from A to D is {(A, C), (C, D)} and that for messages D to A is {(D, C), (C, A)}. Neither of these symmetry characteristics is necessary in Kleinrock's model. However, when they apply, the solution is also symmetric, for example, we will have $C_{11} = C_{12}$, and so on.

---

[c]The numerical values in table 6–1 and others used in this example correspond to those in an example of Kleinrock's with a scale factor that results in more conventional link capacities in bits per second [3].

Table 6-1
Arrival Rates, Messages per Second

From \ To	A	B	C	D	E
A	–	0.935	9.34	0.610	2.94
B	0.935	–	0.820	0.131	0.608
C	9.34	0.820	–	0.628	2.40
D	0.610	0.131	0.628	–	0.753
E	2.94	0.608	2.40	0.753	–

When these symmetries do not apply, the solution is in general not symmetric.[d]

With respect to the choice of routes, Kleinrock points out that equation 6.10 shows that it is generally desirable in minimizing T to keep n, the average number of links per message, low. Thus we will choose as the route between A and E, for example, {(A, B), (B, E)} rather than the route via nodes C and D since the latter is one link longer than the former. Other arbitrary route choices that are not resolved by this rule are A–C–D and C–D–E. None of these route choices is necessary in Kleinrock's model, but with different choices, the solutions would obviously not be the same as we will find here.

By adding all the entries in table 6–1, we find that $\gamma = 38.33$. To arrive at the λ_i, we simply sum for each link the γ_{ij} that traverse it. For example, for link (C, D) we have $\lambda_{11} = \gamma_{AD} + \gamma_{CD} + \gamma_{CE} = 0.610 + 0.628 + 2.40 = 3.638$; by symmetry, this is also the value of λ_{12}, for link (D, C). Continuing in this way, we find $\lambda = \lambda_1 + \lambda_2 + \cdots + \lambda_{14} = 50.23$. Thus $n = \lambda/\gamma = 50.23/38.33 = 1.31$.

We now assume that the total link capacity is $C = C_1 + C_2 + \cdots + C_{14} = 38{,}330$ bits per second, and that the average message length is 100 bits, that is, $\mu = 0.01$. Thus $\rho = \gamma/\mu C = 38.33/(0.01 \times 38{,}330) = 0.1$. Applying equation 6.8, the optimum link capacities in bits per second are

$$C_1 = C_2 = 3{,}130 \quad C_9 = C_{10} = 2{,}990$$
$$C_3 = C_4 = 5{,}390 \quad C_{11} = C_{12} = 3{,}020$$
$$C_5 = C_6 = 1{,}340 \quad C_{13} = C_{14} = 2{,}790$$
$$C_7 = C_8 = 517$$

---

[d] In practical systems it is seldom possible to have different capacities for the two oppositely directed links between a pair of nodes. Also many transmission schemes require that there be acknowledgments sent from receiving node to sending node for messages or for blocks of data so that transmission is not truly unidirectional. The model nevertheless applies reasonably well to a network with duplex links.

Message-Switched Networks

The average time in the system for all messages, using equation 6.10, is $T = 0.045$ seconds. If we were to increase the demands proportionally, for example, by multiplying all the entries in table 6–1 by 6 ($\rho = 0.6$), we would find that T increases to 0.18 seconds. Of course, as ρ approaches $1/n = 1/1.31 = 0.763$, T increases very rapidly.

Extensions

Kleinrock derives the optimum capacity assignment and minimum average time in the system under a more general cost constraint than C equals a constant, for which equations 6.8 and 6.10 apply [3]. The more general form is $D = \Sigma d_i C_i$ equals a constant, where d_i equals the cost of one unit (one bit per second) of capacity on link i. Our case corresponds to $d_i = 1$ for all links, and $D = C$. The optimum capacity for the ith link with the more general cost constraint is

$$C_i = \frac{(\lambda_i/\mu) + (D_e/d_i)(\lambda_i d_i)^{1/2}}{\sum_j (\lambda_j d_j)^{1/2}} \tag{6.11}$$

With this set of link capacities

$$T_i = \frac{\sum_j (\lambda_j d_j)^{1/2}}{\mu D_e (\lambda_i d_i)^{1/2}} \tag{6.12}$$

and

$$T = \frac{n \sum_j (\lambda_i d_i)^2}{\mu D_e \lambda} \tag{6.13}$$

provided that $D_e > 0$, where

$$D_e = D - \frac{\sum_j \lambda_j d_j}{\mu} \tag{6.14}$$

If the slightly more general linear cost constraint $D = \Sigma(d_i C_i + d_{0i})$ equals a constant prevails, where d_{0i} is a fixed cost for each link i, this solution is modified by replacing D in equation 6.14 with $D - \Sigma d_{0i}$.

Kleinrock also considers still more general cost functions that are not linear forms in the C_i, as well as the difficult problem of capacity optimization when the C_i are restricted to a set of fixed values, for example, 50 kilobits per second (kbps) and 250 kbps, rather than having a continuous range of values as in equations 6.8 and 6.11 [2].

The optimization results given so far all apply when the network average time in the system, T, expressed by equation 6.7, is minimized. Meister et al [5] found that in some situations the optimum link capacity solution for this case results in a rather wide variation among the link average times T_i, equation 6.6. As a result, messages between some node pairs experience a much greater average delay than those between other node pairs. They found the optimum link capacity assignments when the alternate quantity

$$T' = \Sigma \left(\frac{T_i^\kappa \lambda_i}{\gamma} \right)^{1/\kappa}$$

is minimized, where $\kappa > 0$; $\kappa = 1$ is the basis for Kleinrock's solution, equation 6.11. They show that when κ has moderately high values, for example, 16 to 64, the T_i with link capacities that minimize T' in a network such as that of figure 6–4 have a much smaller range of variation than they have when the link capacities are assigned so as to minimize T ($\kappa = 1$). This is achieved with only a modest increase in T, the average time in the system for all messages, which remains the performance criterion. Their solution for the optimum link capacities is

$$C_i = \left(\frac{\lambda_i}{\mu} \right) + \frac{\left(\frac{D_e}{d_i} \right) (\lambda_i d_i^\kappa)^{1/(1+\kappa)}}{\sum_j (\lambda_j d_j^\kappa)^{1/(1+\kappa)}} \qquad (6.15)$$

and the network average time in the system is

$$T = \frac{n^{1/\kappa} \left[\sum_j (\lambda_i d_i^\kappa)^{1/(1+\kappa)} \right]^{(1+\kappa)/\kappa}}{\mu D_e \lambda} \qquad (6.16)$$

in which D_e has the same meaning as in equation 6.14. The values of the link times T_i may be found, as usual, by substituting the capacities from equation 6.15 into equation 6.6.

A Dual Problem

We have so far found the allocation of the C_i under various constraints that minimizes T. A dual problem is to prescribe the maximum allowable value of T, say, T_{MAX}, and then find the set of C_i values that minimizes the cost. The solution when the costs are linear in the C_i, that is, when the total cost is $D = \Sigma(d_i C_i + d_{0i})$, follows [2].

$$C_i = (\lambda_i/\mu) + \frac{\lambda_i \Sigma(\lambda_j d_j)^{1/2}}{[\mu \gamma T_{MAX}(\lambda_i d_i)]^{1/2}} \quad (6.17)$$

and

$$D = \Sigma[(d_i \lambda_i/\mu) + d_{0i}] + \frac{[\Sigma(\lambda_i d_i/\mu)^{1/2}]^2}{\gamma T_{MAX}} \quad (6.18)$$

Effects of Network Connectivity, Alternate Routing, and Random Routing

So far we have dealt with a fixed network connectivity or topology. Kleinrock notes two properties of T from equation 6.10 that enter into the comparison of networks of different connectivities [2,3]. First, since T is proportional to n, the average route length, it is desirable to keep this quantity low, as we have noted in connection with the example of figure 6–4. Ideally, this is achieved in a fully connected network in which each node is connected by a link to each other node; then $n = 1$ with the obvious choice of routes. Second, the summation in the numerator on the right side of equation 6.10 is least when one of the λ_i/λ terms is 1 and all the others are zero. Since it is not practical to concentrate the flow in a network this much (in one link), the conclusion is that there should at least be a small number of high-capacity links that carry most of the traffic. These goals conflict with each other, and it turns out that the first (high connectivity) is more important in reducing T when ρ is near 1, whereas the second (high concentration of traffic in a few links) is more important when ρ is near zero.

Kleinrock shows that these considerations apply when there is fixed routing, that is, just one route from node i to node j [3]. To find out the effects of other types of routing, he carried out simulations of networks with a particular kind of *alternate routing*, defined as follows. At each node i there is a table that contains, for messages addressed to node j, an ordered list of links connected to i. If a message arrives at i for j, whether from outside the backbone network, that is, from a user terminal (see figure 6–1) or from another node, the switching equipment at i transmits this message on the first link in the list for j if this link is free; on the second link in the list, if the first link is busy; and so forth. If all the links in the list are busy, the message is held in a queue at i until one of them becomes free. Referring to figure 6–1, for example, if the ordered list of links at node $i = 4$ for messages to node $j = 1$ is [(4, 2), (4, 3)], then a message arriving at 4 addressed to 1 will be transmitted on link (4, 2) if this link is immediately available; on link (4, 3) if it is not; and on the first of these two links to become free if both are busy at the time the node 4 switching equipment first tries to send the message. If we assume, for example, that the message is sent from 4 to 3, then the switching equipment at the latter node consults a similar table that gives its ordered choices of links for the further routing of the message. The first-choice link at 3 would be most likely be (3, 1). Thus the message progresses from one node to the next until it reaches its destination. Kleinrock's simulations showed that, when the γ_{ij} are known and do not vary with time, this kind of alternate routing does not reduce the value of T with optimum capacities compared to its value with fixed routing, although it may serve to reduce T when the capacities are nonoptimum. [The effects of time variations in the γ_{ij} are especially significant in packet-switched networks, which we will consider in chapter 7. In these networks it is common to employ *adaptive routing*, in which the route for a message (packet) is made to vary from time to time depending on such network conditions as the current link loadings, λ_i. Kleinrock's delay model does not apply to the case of adaptive routing.]

Kleinrock also applied his delay model, embodied in equations 6.6 and 6.7, to certain networks with *random routing*. Here the choice of the next link at node i for sending a message to node j, assuming there are $K_i > 1$ links that emanate from i, as given by a set of probabilities $p_m(j), m = 1, 2, \ldots, K_i$ with $\Sigma p_m(j) = 1$. This kind of routing, since it uses no information about the network connectivity, tends to increase greatly the average number of links per message, n, and hence the average time in the system, T; the message arrives at its destination literally by chance. Kleinrock compared the analytic results for networks with random routing with simulation results for networks with fixed routing, and confirmed these conclusions. The benefit of random routing is that a message may continue to be routed through the network even if there are drastic changes in the network, for example, destruction of links, with no dissemination of network status information being necessary.

6–5 Network Optimization:
The Flow Assignment Problem

A quite different problem from that of capacity assignment is that of *flow assignment* (FA), also known as *routing optimization*. Here the network

Message-Switched Networks

graph, the link capacities, and the message arrival rates are given, and we wish to find the routing for messages between each node pair that minimizes T. As Kleinrock points out, we have no assurance a priori in this case that it is feasible to have only one route between any pair of nodes i and j since it may happen that $\gamma_{ij}/\mu > C_k$, where C_k is the capacity of some link that is a bottleneck for such messages, for example, the smallest of the link capacities in an otherwise desirable route from i to j [2]. Even if this constraint does not occur, it may be impossible to route γ_{ij} as well as other node-to-node throughput components over single routes. [The feasibility of the solution to the problem of section 6–3 is assured by the positiveness of D_e, equation 6.14, in the general case, and by the implicitly satisfied criteria $n\rho < 1$ and $C_i > \lambda_i/\mu$ in the case of the more limited cost constraint C equals a constant.] Thus unlike the situation in section 6–4, we may require two or more routes from i to j.

The solution of the routing optimization problem requires in turn the solution of a *multicommodity flow* problem, as follows. Assume that the links of the network are numbered from 1 to L, and the nodes from 1 to N. We will refer to $\gamma_{mn}/\mu \geq 0$, bits per second, as the total *flow* from node m to node n ($m, n = 1, 2, \ldots, N$; $m \neq n$). Let the proportion of this flow that exists on link k be $r_k^{(m,n)}$ ($0 \leq r_k \leq 1$; $k = 1, 2, \ldots, L$). We denote by $f_k^{(m,n)} = r_k^{(m,n)} \gamma_{mn}/\mu$ the flow in link k that results from messages directed from node m to node n; we refer to a directed graph, as in section 4–3. Then, in analogy to equation 4.1, the flows $f_k^{(m,n)}$ must satisfy at each node i for each node pair m, n the conservation of flow equation

$$\underset{\substack{\text{Links directed} \\ \text{outward at} \\ \text{node } i}}{\sum} f_i^{(m,n)} - \underset{\substack{\text{Links directed} \\ \text{inward toward} \\ \text{node } i}}{\sum} f_i^{(m,n)} = \begin{cases} \gamma_{mn}/\mu, & \text{if } i = m \\ 0, & \text{if } i \neq m \text{ or } n \\ -\gamma_{mn}/\mu, & \text{if } i = n \end{cases}$$

(6.19)

As implied in the equation, it applies only to links incident at node i, that is, those for which one end is at node i. Furthermore, in analogy to equation 4.2, the total flow in each link, that is $F_k = \Sigma_{m,n} f_k^{(m,n)}$ may not exceed its capacity

$$C_k \geq F_k \tag{6.20}$$

(The nonnegative nature of each individual flow $f_k^{(m,n)}$ is assured by its definition.) We now observe that our previously defined total link message rate is just $\lambda_k = \mu F_k$. Substituting this in equations 6.6 and 6.7, we have for the average time in the system for all messages

$$T = (1/\gamma) \sum_k \frac{F_k}{(C_k - F_k)} \tag{6.21}$$

Finally, we may state the multicommodity flow problem to be solved as follows: it is to select for each commodity γ_{mn}/μ a feasible set of link flows $f_k^{(m,n)}$, that is, a set that obeys equations 6.19 and 6.20, that minimizes T. In order to apply Kleinrock's delay model, as in equation 6.21, the independence assumption, discussed in section 6–3, is adopted for multiple routing, as it was for fixed routing. The solution of this problem shows in theory how to minimize T by selecting the link flows F_i, but it does not generate a routing procedure for the switching nodes that actually achieves these flows [2].

The solution is complex since the general multicommodity flow problem with a nonlinear *objective function* (here T) cannot be solved in closed form but only by iterative approximations. There is an excellent discussion of one approach, the *flow deviation* method, in Kleinrock's book [2]. An alternate method, the *gradient projection* method, is described by Schwartz and Cheung [6].

A still more general problem than either CA or FA is that of *capacity and flow assignment* (CFA). Here, only the network graph and message-arrival rates are given, and we wish to find both the link capacities and the routing that minimize T, subject to a cost constraint such as $D = \Sigma \, d_i C_i =$ constant. Finally, the most general problem is that of topology, capacity, and flow assignment (TCFA), in which only the message-arrival rates (and presumably the node locations) are given, and we must find the network graph, link capacities, and routing that minimize T, under a constraint such as D equaling a constant. These two problems have not been solved, but there are heuristic approaches that give locally optimum solutions, that is, solutions that are founded on a particular feasible flow or capacity assignment but that cannot be proven to be optimum among all possible solutions [2, 7].

References

1. Syski, R. *Introduction to Congestion Theory in Telephone Systems*. Edinburgh: Oliver and Boyd, 1960.
2. Kleinrock, L. *Queueing Systems, Vol. II: Computer Applications*. New York: Wiley, 1976.
3. Kleinrock, L. *Communication Nets, Stochastic Message Flow and Delay*. New York: Dover, 1972 (reprint of 1964 edition).
4. Burke, P.J. "The Output of a Queueing System," *Operations Res.* 4:699–704, 1956.
5. Meister, B., Müller, H.R., and Rudin, H.R., Jr. "New Optimization Criteria for Message-switching Networks," *IEEE Trans. Commun.* COM-19:256–260, 1971.

6. Schwartz, M., and Cheung, C.K. "The Gradient Projection Algorithm for Multiple Routing in Message-switched Networks," *IEEE Trans. Commun.* COM-24:449–456, 1976.
7. Gerla, M., and Kleinrock, L. "On the Topological Design of Distributed Computer Networks," *IEEE Trans. Commun.* COM-25:48–60, 1977.

7 Packet-Switched Networks

7–1 Introduction

Unlike message switching (see chapter 6), which has a history going back forty years or so to torn-tape telegraph switching systems, packet switching was first proposed much more recently by Baran [1]. His paper describes the basic idea of packet switching, which is to modify a message-switched network by dividing each message into *packets* (Baran used the term "blocks") of fixed length and to route each packet from its origin switching node to its destination switching node independently. Subsequently, packet switching was embodied in the ARPANET [2], operated for a U.S. government agency, and it has also been applied in commercial networks in this country and elsewhere.

A packet-switched network, as contrasted with a message-switched network, can handle several different types of traffic concurrently. These include *high-throughput* traffic, for example, the transmission of large data files between computers, for which accuracy and high average data speed are the most important performance requirements; *low-delay* traffic, for example, interactive communication between a person at a terminal and a remote computer, for which accuracy and low average message delay are important; and *real-time* traffic, for example, packetized speech, for which the performance of a circuit-switched connection (see chapter 5) must be approached by maintaining a relatively constant data speed but for which extreme accuracy is not important owing to the redundancy of the information [3]. In all three of these applications, packet switching shares with message switching the advantage over circuit switching that, since a packet is stored temporarily at each node, its format, code, or speed may be changed before it is retransmitted. The first two types of traffic are common in practical packet-switched networks whereas packetized speech is still under study.

In this chapter we examine the extensions of Kleinrock's model of a message-switched network, discussed in chapter 6, to the case of packet switching. Then we examine the methods of network optimization that have been applied to packet-switched networks, which differ in practice from those for message-switched networks discussed in chapter 6, although they are based on some of the same principles.

7-2 Extensions of Kleinrock's Model

Kleinrock's model of a message-switched network, described in section 6–3, can be extended to apply to packet-switched networks [4, 5].

The Single-packet Case

We consider first the estimation of the *average time in the system* for a message that consists of a single packet. As in section 6–3, the time that a packet spends in the system has two principal components: the *service time*, which is the sum of the times it takes to transmit the packet on each link it traverses in the backbone network;[a] and the *delay time*, which is the sum of the delays that the packet experiences at each node waiting for a link to become available. As in section 6–3, we will estimate these two times by looking at one link or node at a time.

If the average packet length is $1/\mu$ bits and the capacity or speed of link i is C_i bits per second, the packet-transmission time or service time on this link is $1/\mu C_i$ seconds.

To estimate the delay time at a node for a packet that is transmitted on link i, we retain the assumptions of the single-server queueing system used in section 6–2, viz., Poisson arrivals of packets and a negative exponential distribution of packet lengths. Then the average packet delay is $W_i = (\lambda_i/\mu C_i)/(\mu C_i - \lambda_i)$ where now λ_i is the arrival rate of packets at link i in packets per second.[b] We must modify this formula in a packet-switched network because there are a significant number of control or overhead packets transmitted such as those for acknowledgments of correct receipt of data packets, that tend to be shorter than data packets. To recognize this, we denote by $1/\mu'$ the average length of all packets, both control and data. Also we denote by λ'_i the total arrival rate of all packets, both control and data, at link i. If we now assume that the single-server queue assumptions apply to the mixed stream of control and data packets, then the average delay for all packets traversing link i is

---

[a] As in chapter 6 we will concentrate here on the backbone or internodal network. In a packet-switched network each user, for example, a terminal or computer, must be connected to a nearby switching node, just as in the case of message switching, by a local-access network. Among the functions of the switching node are the division of messages into packets and the reassembly of incoming packets into messages.

[b] In practical networks the lengths of full packets are constant rather than having a negative exponential distribution. Thus the single-server queue results of section 6–2 do not apply. Although some analyses of packet delays use the queueing formulas appropriate to the M/D/1 queueing system, that is, Poisson arrivals and constant (deterministic) packet lengths, it is customary to use the results for the M/M/1 queueing system, as we do here. A possible justification for this is that the average delays with exponential packet lengths are greater than those for constant packet lengths, and the results are in this sense conservative.

Packet-Switched Networks

$$W'_i = \frac{(\lambda'_i/\mu'C_i)}{(\mu'C_i - \lambda'_i)} \tag{7.1}$$

For packet switching we must also include as components of the time in the system the link propagation time and the nodal processing time, both of which tend to be greater relative to the service time than in the case of message switching. We will assume that the former is P_i seconds for link i and the latter is K seconds per node.

If we now add together the service time, delay time, link propagation time, and nodal processing time, we have as the average time in the system for a single-packet message

$$T_{SP} = K + \sum_i \left(\frac{\lambda_i}{\gamma}\right)\left[\left(\frac{1}{\mu C_i}\right) + W'_i + P_i + K\right] \tag{7.2}$$

Here γ is the total arrival rate of data packets into the backbone network; W'_i is found from equation 7.1; and the extra delay K outside the summation accounts for each packet's traversal of a number of nodes that is one greater than the number of links it traverses. The sum in equation 7.2 extends from $i = 1$ to $i = M$, the number of links in the network. Equation 7.2 retains the independence assumption used in Kleinrock's model, section 6–3, as to both control and data packets, as well as the assumption of fixed routing.

Multiple-packet messages

If on the average a message contains $m > 1$ packets, its average time in the system, that is, in the backbone network, is the sum of three components. The first of these is average time in the system for the first packet, which is given by equation 7.2 with the term $1/\mu_F C_i$ substituted for $1/\mu C_i$, $1/\mu_F$ being the length of a full data packet in bits. (In general $1/\mu_F > 1/\mu$ since the latter is the average length of all data packets, including those of less than full length.)

The second component is the average link-transmission time for the remaining $m - 1$ packets, that is, for $m - 2$ packets of full length and one packet, the last in the message, of average length. The appropriate average transmission time is thus

$$\sum_i \left(\frac{\lambda_i}{\lambda}\right)\left[\left(\frac{(m-2)}{\mu_F C_i}\right) + \left(\frac{1}{\mu C_i}\right)\right]$$

Here $\lambda = \Sigma \lambda_{ij}$ is the network throughput in data packets per second.

The third component is of a different nature from the other two. It is the average *interpacket gap time*, that is, the sum of the $m - 1$ intervals between the m packets of the message. Owing to mathematical difficulties in calculating this component explicitly, it is approximated as

$$(m - 1)\rho(1 - \rho^{n-1})S_F$$

where ρ is the average link utilization, that is $\rho = \Sigma_i(\lambda'_i/\mu'C_i)/M$; n is the average number of links traversed by a message, that is, $n = \lambda/\gamma$; and S_F is the transmission time for a full packet, averaged over all links, that is,

$$S_F = \frac{\sum_i(\lambda_i/\lambda)}{\mu_F C_i}$$

By adding together the three components, we have as the average time of a multipacket message in the backbone network[c]

$$T_{MP} = K + \sum_i \left(\frac{\lambda_i}{\gamma}\right)\left[\left(\frac{1}{\mu_F C_i}\right) + W'_i + P_i + K\right]$$

$$+ \sum_i \left(\frac{\lambda_i}{\lambda}\right)\left[\left(\frac{m-2}{\mu_F C_i}\right) + \left(\frac{1}{\mu C_i}\right)\right] + (m-1)\rho(1-\rho^{n-1})S_F \quad (7.3)$$

Some Numerical Values

Equation 7.2 for the average time in the system of a single-packet message and equation 7.3 for the multipacket message are quite complex. To illustrate the properties of one packet-switched network, we draw on data for the ARPANET. In anaylzing some observations on this network, Kleinrock and Naylor found that the average time in the system was not much different for very low values of ρ, the average link utilization, than for higher values of ρ typical of normal traffic loads [4]. Therefore, they concluded, the W'_i terms in equations 7.2 and 7.3 may be neglected. If we assume further that the link capacities are all the same, that is, that $C_i = C$, then equation 7.2 reduces to

---

[c]Equation 7.3 differs slightly from the corresponding equations in references 4 and 5 because we retain the assumption that the processing time is the same at all nodes, whereas in these references this processing time is node dependent. Also references 4 and 5 include a term for transmission time of a packet from the node processor to a connected host computer, which is not attributable to the backbone network.

Packet-Switched Networks

$$T_{SP} \approx (n + 1)K + \left(\frac{n}{\mu C}\right) + \sum_i \left(\frac{\lambda_i}{\gamma}\right) P_i \qquad (7.4)$$

The term C obviously depends on the particular network. In a practical network of large geographic extent, n may have a value in the range of 1 to 10; Kleinrock and Naylor found $n = 3.31$ for ARPANET. The value of K depends primarily on the speed of the nodal processor; the ARPANET value was 0.75 milliseconds (ms). As the average packet transmission time $1/\mu C$ with $C = 50$ kbps, Kleinrock and Naylor found 8.2 ms, indicating an average packet length of $1/\mu = 410$ bits. To find the third term on the right side of equation 7.4, it is necessary to know the link-arrival rates λ_i and the network throughput γ, which are specific network characteristics, and the link-propagation times P_i, which are properties of the links and their lengths. For the observed ARPANET layout, this term was 11.4 ms. Thus $T_{SP} \approx (3.31 + 1) \times 0.75 + (3.31 \times 8.2) + 11.4 = 41.0$ ms, of which the link-transmission time $n/\mu C = 27.1$ ms is the largest part.

Round-trip Delay

Of the three types of traffic described in section 7-1, low-delay or interactive traffic has so far predominated in ARPANET and in commercial packet-switched networks such as those operated by GTE Telenet and Tymnet in the United States. For this reason, there is great interest in the *average round-trip delay*, T_R, that is, the sum of the average time in the system for a message and the average time for a reply message to return from the destination node. (From the viewpoint of a human network user, if T_R exceeds about one-half second in interactive communication, the system seems annoyingly slow.) Kleinrock and Naylor estimated the first component of T_R for ARPANET as simply T_{SP}, the average time in the system for a single-packet message, since, in the observations they analyzed, 96 percent of all messages consisted of single packets [4]. They further estimated the second component of T_R as the average time in the system for a particular kind of reply message in ARPANET, viz., and acknowledgment of correct receipt of a packet called a Request for Next Message (RFNM). A RFNM is a single-packet message whose transmission time at 50 kbps was 3.36 ms, indicating an average length of 168 bits. Thus, if we apply equation 7.4, we find an average time in the system for a RFNM of 25.7 ms. Adding this to the average time in the system for a single-packet (data) message, as calculated earlier, we find an estimated round-trip delay of $T_R = 25.7 + 41.0 = 66.7$ ms. (Note that a crude estimate of T_R under the assumptions here is just $2T_{SP}$.) Kleinrock and Naylor report a slightly different estimate, 73 ms, since their calculations take into account more specifically the properties of RFNM traffic as well as delays in the nonbackbone feeder network. The observed value of T_R for one week of ARPANET experience was higher, 93 ms, which they attribute primarily to additional host-computer delays not comprehended in our model and to peaking of ρ, the average link utilization.

Threshold Network Throughput

Kleinrock suggests that the average delay in a packet-switched network is approximately constant versus the traffic load γ, until γ approaches a threshold value, γ^* [5]. This is the smallest value of γ for which $\gamma_m = \mu C_m$ for some link m. That is, link m is the critical link or bottleneck that approaches a high-delay condition. [See section 6–4. A somewhat more general condition than that of the saturated bottleneck link is that of a saturated cut (see section 4–3).] This is strictly true only with fixed routing, and we will see in section 7–3 that practical packet-switched networks tend to have adaptive routing. Nevertheless, simulations of the delay versus γ for ARPANET show that γ^* can be found reasonably well in this way [5]. The delay below γ^* is approximated by equations such as equation 7.4; note that the delay terms are absent in this equation, and since the ratios λ_i/γ do not vary with γ for a fixed routing pattern, neither does the value of T_{SP}.

7–3 Network Optimization

In our discussion of message-switched networks in chapter 6, we identified several key problems such as those of *capacity assignment* (CA) and *flow assignment* (FA); the latter, we noted, is the same as *routing optimization*. These problems have been approached in quite different ways for packet-switched networks.

Capacity Assignment

Here the network graph and the node-to-node packet-arrival rates are given, and we wish to select the link capacities C_i so as to minimize a delay measure, for example, the single-packet delay T_{SP} given by equation 7.2. A solution such as Kleinrock's CA procedure for the message-switched network would apply only if the total arrival rate of packets γ and its node-to-node components γ_{ij} were known and invariant with time. Also the specific CA solution in section 6–4 requires the assumption of single, fixed routes between node pairs. In practical packet-switched networks, in contrast, γ

Packet-Switched Networks

and γ_{ij} are usually known only approximately and tend to change with time. Thus these networks tend to use routing schemes that are *adaptive* in the sense that a packet going from node i to node j does not necessarily follow the same route at one time as at another, depending on the status of the network. As we will see later in this section, this kind of routing allows the network to react not only to variations in the node-to-node submitted traffic but also to failures of links or nodes.

In view of the inapplicability of the optimal CA method, link capacities in practical packet-switched networks are assigned by simpler methods. One such method is to assign each node-to-node capacity requirement, that is, the product of the packet-arrival rate and the average packet length in bits, or γ_{ij}/μ, to the links of a route that has ostensibly low delay, for example, one that contains the fewest intermediate nodes of all possible routes between the two nodes in question. The capacity of each link is then set at the available discrete capacity that is closest to the sum of all the node-to-node capacity requirements that are thus assigned to the link. (Typically link capacities are chosen from a small set of discrete values, for example, 9.6 kbps, 56 kbps, and 200 kbps.) Since packet-switched networks can afford high reliability, owing to their ability to route packets around a failed link or node, the backbone network is often designed to have a minimum connectivity, for example, a minimum of two or three links terminating at each node (see section 4-4). Thus if some of the link capacities derived by the procedure just described are too high, some node-pair capacity requirements may be diverted altogether to different routes or divided over two or more routes. In brief, link capacities are assigned to meet the packet transmission-time requirements and the network-reliability requirements rather than according to a performance-optimization criterion.

Flow Assignment

In this problem, as noted in chapter 6, the network graph, link capacities, and node-to-node packet-arrival rates are given, and we wish to find a method of routing packets for each node pair that optimizes a network-performance measure, for example, one that minimizes the average packet delay.

Certain theoretical methods of flow assignment in message-switched networks have been applied to packet-switched networks, for example, the flow deviation method [5] mentioned in section 6-5 and the extremal flows method [6]. However, these methods are based on solutions of the multicommodity-flow problem in which the *objective function*, or quantity to be minimized, is the average time in the system for all packets, using only the delay term as given in equations 6.6 and 6.7, viz, $T = \Sigma_i \lambda_i/\gamma(\mu C_i - \lambda_i)$; the interpretations of the symbols here are as in section 7-2. We have noted that the delay term seems in practical packet-switched networks not to be the controlling element in such performance measures as the round-trip packet delay. Furthermore, the multicommodity-flow solutions do not apply to networks with adaptive routing. Thus, although these solutions to the FA problem are of great interest in showing the theoretical limitations of packet-switched networks under heavy traffic loading, they do not correspond directly to the methods of routing used in practical networks.

To achieve routing optimization in packet-switched networks, it is presently the practice to select routes adaptively on the basis of information about the state of the network that is in some systems not older than a few seconds[c] and is in other systems updated less frequently, for example, only when a link or node fails [7]. This information consists primarily of the "costs" or weights of the links. The most desirable route for sending a packet from one node to another is that of least cost, or in other words, a shortest route, as we have used this term in chapter 3, since the link costs add to make up the route cost. For example, if the cost of a link is the current average packet delay on this link, including transmission time, propagation time, and so on, then the selected route is that which has the least estimated overall packet delay between the nodes in question. Alternatively, the link cost may be a measure of the current or recent utilization of the link. If highly utilized links are given greater costs than those that are relatively underutilized, packets will be diverted to the latter class of links.

Other link costs are also used that do not change with time. For example, when costs are based on link speed, it is generally desirable to get as many packets on high-speed links as possible; thus a link at 2,400 bps is given a higher cost than one at 9,600 bps. As another example, packets of interactive traffic can be diverted from communications satellite links, with their inherently long propagation time, by making the cost of such links high for this type of packet but low for other types, for example, packets of a long file transfer.

There are major design problems to select the types of costs and assign their values; to assure that the least-cost (shortest-route) algorithm is highly efficient since it is executed often; and to control the dissemination of link cost data throughout the network. These problems and others, along with descriptions of routing techniques in several present packet-switched networks, are discussed in a recent paper by Schwartz and Stern [7]. They also discuss two shortest-route algorithms that are widely used: that of Dijkstra (see section 3–4) and another due to Ford and Fulkerson [8].

It is interesting that in most of the networks surveyed by Schwartz and Stern, separate packets of the same message may not be routed differently

---

[c]If the interval for updating the network-status information is made extremely short, the network may loose efficiency. For example, too much link capacity may be used in disseminating change-of-status data to all nodes. Another hazard is that of instability in routing patterns.

from one another but rather always take the same route. (The packets of a later message between the same nodes may take a different route.) Thus the simplification of routing processing at the nodes has so far taken precedence over exacting the greatest measure of routing diversity, and potentially of network reliability. As Schwartz and Stern point out, the use of a fixed route for all packets of a message also avoids the problem of these packets' arriving at the destination node out of order.

Flow Control

As a result of the fluctuations of the γ_{ij} with time, peculiarities of the routing algorithm, temporary full occupancy of buffers for storing packets at the nodes, and other dynamic effects in packet-switched networks, it may be necessary at times to impose *flow control*. That is, the current γ_{ij} must be reduced to avoid a spread of congestion throughout the network and a consequent reduction in throughput. Two principal types of flow-control techniques are those in which node pairs are limited in their traffic, that is, in which the γ_{ij} are selectively or uniformly limited; and those in which a networkwide limit is placed on the total number of packets that may be present. The second type is evidently centralized whereas the first can be carried out by individual nodes [9].

References

1. Baran, P. "On Distributed Communication Networks," *IEEE Trans. Commun. Systems* COM-12:1–9, 1964.
2. Roberts, L.G., and Wessler, B.D. "Computer Network Development to Achieve Resource Sharing," *Proc. AFIPS Spring Joint Comput. Conf.* 36:543–549, 1970.
3. Kimbleton, S.R., and Schneider, G.M. "Computer Communication Networks: Approaches, Objectives, and Performance Considerations," *Comput. Surv.* (US) 7:129–173, 1975.
4. Kleinrock, L., and Naylor, W.E. "On Measured Behavior of the ARPA Network," *Proc. AFIPS Nat. Comput. Conf.* 43:767–780, 1974.
5. Kleinrock, L. *Queueing Systems, Vol. II: Computer Applications.* New York: Wiley, 1976.
6. Cantor, D.G., and Gerla, M. "Optimal Routing in Packet-switched Computer Networks," *IEEE Trans. Computers* C-23:1062–1069, 1974.
7. Schwartz, M., and Stern, T.E. "Routing Techniques Used in Computer

Communication Networks," *IEEE Trans. Commun.* COM-28:539–552, 1980.
8. Ford, L.R., Jr., and Fulkerson, D.R. *Flows in Networks*. Princeton, N.J.: Princeton University Press, 1962;.
9. Schwartz, M. *Computer-Communication Network Design and Analysis*, Englewood Cliffs, N.J.: Prentice-Hall, 1977.

8 Centralized Computer Networks

8-1 Introduction

It is common to connect remote-user terminals to a computer in networks such as those shown in figure 8–1. In figure 8–1(a), the terminals are on *multipoint* or *multidrop* private lines that terminate at the computer. In the *hierarchical* network of figure 8–1(b), the terminals are on *point-to-point* lines that terminate at *multiplexers* or *concentrators*, which in turn are connected to the computer with similar lines; the lines that connect terminals to multiplexers/concentrators are usually of lower speed than those that connect the latter devices to the computer.[a] There are may other possible configurations, the choice depending on the amount of traffic per terminal, line costs, required reliability, and so on. For example, the multiplexors/concentrators in a hierarchical network may themselves be on multipoint lines. The primary advantage of the hierarchical network is its economy, especially when the terminals are clustered in areas that are some distance from the computer, since the total costs for multiplexers/concentrators, high-speed lines, and low-speed lines in figure 8–1(b) are often less than those for low-speed lines in figure 8–1(a).

We have dealt with the design of minimum-cost multipoint lines as trees or segments of trees in chapter 2. These results are directly applicable to networks of the kind shown in figure 8–1(a). In this chapter, we examine two related problems that are of practical interest in centralized computer networks. The first of these is the estimation of response time, and the second is the selection of multiplexer/concentrator locations for a hierarchical network.

8-2 Estimation of Response Time

When the user enters a message at a terminal in figure 8–1(a), there are several time intervals that elapse, in sequence, before he receives the

---

[a] A multiplexer interleaves the signals from the connected terminals on the high-speed line; thus the speed of this line is approximately the sum of the operating speeds of the terminals. A concentrator, on the other hand, switches the signals from the terminals to the high-speed line, whose speed is generally less than the sum of the operating speeds of the terminals. Thus the concentrator relies for its operation on the intermittent or random nature of messages from the terminals whereas the mulitplexer does not.

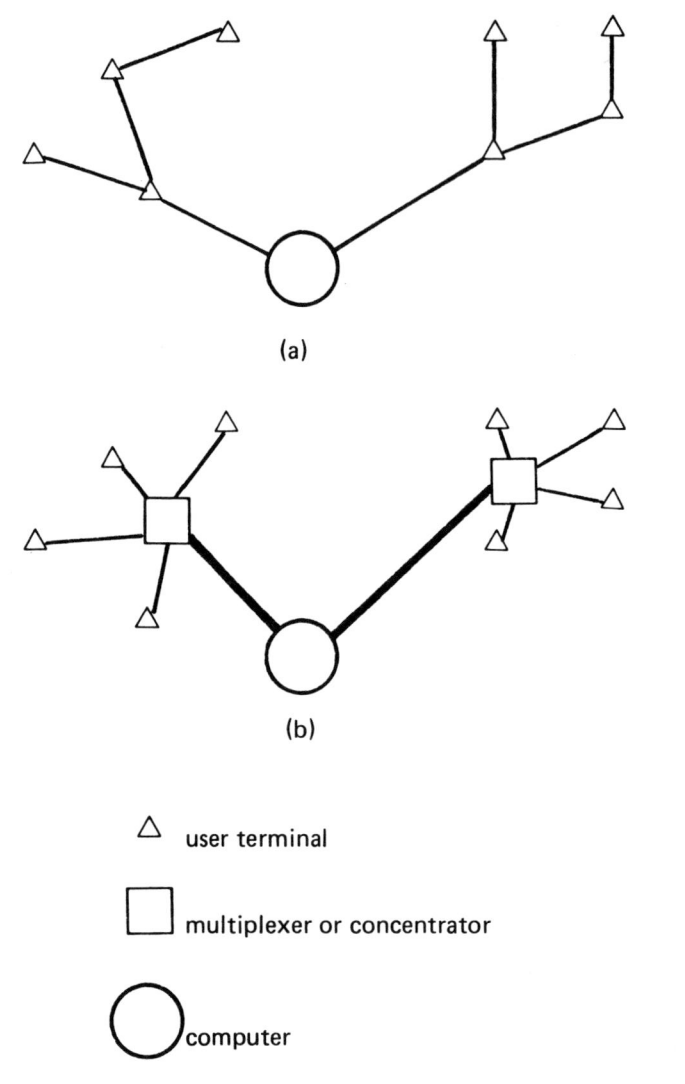

Figure 8–1. (*a*) Nonhierarchical Network (*b*) Hierarchical Network

computer's response, for example, the display or printing of a message at the terminal. We will refer to the sum of these intervals as the *response time*. This is the major performance measure, especially in networks for interactive communication.

First, if the terminal is one of several connected to a local controller, the latter device may already have stored messages from other terminals or it

Centralized Computer Networks 115

may be occupied with a message from the computer, and the user's message may thus wait in a local queue until the controller is free to act on it. Second, since several of the terminals or controllers in figure 8–1(a) share the use of the communications line, there is a queueing delay to obtain the use of the line; its length depends on the amount of terminal activity, the line speed, and the *line discipline*. By the latter term, we mean the method by which a terminal or controller bids for the use of the line, obtains this use, sends one or more messages, and signals the computer that it no longer requires the line. We will consider later in more detail two types of line discipline, *contention* and *polling*.

The third component of response time is the propagation time for signals between the terminal and the computer; this applies also to the return message. The fourth component is the transmission time, b/C seconds, where b is the length of the message in bits, and C is the line capacity or speed in bits per second.[b] Fifth, and finally, there is the time that the computer needs to receive (and perhaps to store and retrieve) the message, to process it, and to generate its reply.

In our analysis we will assume that the first and third components, viz., the local queueing delays and the propagation times, are negligibly small compared to the other response-time components. Thus we will deal with the queueing delay for the use of the line, the transmission time, and the computer-service time.

A Contention Model

We consider a single multipoint line that connects several terminals to the computer. Suppose that the terminals originate messages at random, that is, that the arrivals of messages at the terminals have a Poisson distribution. (See section 6–2. We assume a stable busy hour, and so on.) Assume that when a terminal originates a message, it may immediately seize the line, send its message, and receive the computer-response message unless the line is occupied either with another terminal's message or its computer-response message; if the line is thus in use, the newly arriving message must wait in a queue until the line is available.[c] We have now described a particular line discipline: a terminal holds the line not only for the duration of its message but also for the duration of the computer's response message, which we assume is sent as soon as the computer can do so and in particular before it

---

[b]We will apply this formula to an input or terminal message, to an output or computer message, or to both, as is required in various situations.

[c]We do not need to define the mechanism by which this queue, consisting in general of messages at various terminals, is maintained. This will become clearer when we discuss polling methods.

may process any other terminal message.[d] Our estimate of the response time, which is the interval between the terminal's seizure of the line and the beginning of the receipt of the response message, thus applies only for this discipline.

There is an evident parallel between this situation and the single-server queue of section 6–2; the multipoint line is the single server for the terminal messages and their respective computer-response messages [1]. The *service time* for the multipoint line is the sum of three intervals: the transmission time for a terminal message, the computer-service time, and the transmission time for a computer-response message. Let us denote the average service time by t_s; in a self explanatory notation, we have

$$t_s = t_{tm} + t_{cs} + t_{cr} \tag{8.1}$$

Thus t_{tm} is the average terminal message-transmission time, and so on. We will assume that the service times have the same Poisson distribution of arrivals as do the terminal messages. If the service times also had a negative exponential distribution of lengths, we could apply the single-server queue model of section 6–2 to find the average response time, that is, the sum of t_s and the average delay that a terminal message experiences in queueing for the line. However, we choose not to assume in general that the service times have such a distribution of lengths. Instead, we will assume that, whatever the actual distribution may be, only the average quantities t_{tm}, t_{cs}, and t_{cr} and the variances of the terminal message-transmission time, computer-service time, and computer-response time, which we will call v_{tm}, v_{cs}, and v_{cr} are known. Thus, assuming statistical independence of the quantities, the variance of the service times is

$$v_s = v_{tm} + v_{cs} + v_{cr} \tag{8.2}$$

We may now apply a remarkable result of queueing theory, the Pollaczek-Khintchine formula [2], which gives the average queueing delay for a single-server system with Poisson arrivals and an arbitrary distribution of service times as

$$\tau = \rho t_s \left[1 + \left(\frac{v_s}{t_s^2} \right) \right] / 2(1 - \rho) \tag{8.3}$$

---

[d]This discipline is suited for a half-duplex line, on which commutication may take place in only one direction (from terminal to computer or vice versa) at a time, but it could also be used with a duplex line.

Here $\rho = \lambda t_s$ is the *utilization factor* of the multipoint line; λ is the arrival rate of messages from all terminals, in messages per second.[e] The response time is $T = t_s + \tau$, which corresponds to equation 6.4.

To illustrate the use of equation 8.3, suppose that we have 10 terminals with an average message-arrival rate of 90 messages per terminal per hour; thus $\lambda = 10 \times 90/3{,}600 = 0.25$ messages per second. Assume that these messages have a constant length of 800 bits and that the line speed is $C = 2{,}400$ bits per second (bps); thus $t_{tm} = 800/2{,}400 = 0.333$ seconds, and $v_{tm} = 0$. If we assume further that the average computer-service time is 0.2 seconds with a variance of 0.1 seconds-squared, and that the average computer-response message length is 1,200 bits with a variance of 4.608×10^6 bits-squared (thus $t_{cr} = 0.5$ seconds and $v_{cr} = 0.8$ seconds-squared), we find $t_s = 0.333 + 0.2 + 0.5 = 1.033$ seconds and $v_s = 0.1 + 0.8 = 0.9$ seconds-squared. Also $\rho = \lambda t_s = 0.25 \times 1.033 = 0.258$. Substituting in equation 8.3 we have for the average delay

$$\tau = 0.258 \times 1.033 \times \frac{[1 + (0.9/1.033^2)]}{2 \times (1 - 0.258)}$$
$$= 0.331 \text{ seconds}$$

The average response time is $T = t_s + \tau = 1.022 + 0.331 = 1.364$ seconds. Since the utilization factor ρ is small here, the delay component of T is small compared to t_s. If we increase ρ by increasing the message-arrival rate to 270 messages per terminal per hour ($\rho = 0.775$) we find that $\tau = 3.280$ seconds and $T = 4.313$ seconds; the queueing delay is much greater. Finally, we note that as ρ approaches 1 in equation 8.3, or in this example as λ approaches $1/t_s = 0.968$ messages per second, τ increases without bound.

This model can be adapted to other line disciplines. For example, if the line is arranged for duplex communication and computer-response messages are transmitted in one direction at the same time that terminal messages are transmitted in the opposite direction, there are two separate queues, one for each direction, and correspondingly, two separate response times. The average delay times for each of the queues may be estimated with equation 8.3, provided that one makes the necessary changes in the arguments. A different solution applies if the line discipline, referring again to half-duplex communication, gives output (computer) messages priority over input (ter-

---

[e]This definition of ρ is consistent with that of equation 6.3 since the average service time t_s is exactly analogous to the quantity $1/\mu C$ of section 6–2. For a negative exponential distribution of service times, $v_s = t_s^2$ and equation 8.3 reduces to $\tau = \rho t_s/(1 - \rho)$ which is equivalent to equation 6.2. The Pollaczek-Khintchine formula applies irrespective of the queue discipline. As is the case with the single-server model of section 6–2, the formula applies when the queue may be of arbitrarily large size.

minal) messages and if, after a terminal has sent its message, it does not hold the line; the line may be seized by another terminal, but only if there is no computer message waiting. In this case, equation 8.3 does not apply, and more complex expressions for the delays for the two priority classes must be used. The solutions for this and other types of line discipline are described in detail by Martin [1].

In our illustrative example and in those for other line disciplines described by Martin, we can generally estimate the variance of the queueing delay, v_d. Thus we can find not only the mean response time $T = t_s + \tau$ but also the variance of the response time $v_r = v_s + v_d$. If we now assume that the response times have a particular kind of two-parameter distribution, then we may draw conclusions about the stastistics of the response times. For example, we may estimate the proportion of response times that do not exceed some given value, t_1. These methods are also described by Martin [1]. Owing to the uncertainties that often exist in the input data, for example, in such quantities as v_{tm}, and so on, the author has found these calculations to be of limited use. In many networks a good view of the performance can be had simply from the line utilization, ρ.[f] If ρ is low, say, 0.5 or less, there are not likely to be serious message delays in a network that is otherwise operating properly; if ρ is in the range of 0.5 to 0.75, there is reason for caution to be sure that temporary overloads do not occur; and it can be quite risky to operate a network with ρ greater than 0.75.

Polling

In practical nonhierarchical networks, the modem at the computer end of each multipoint line or segment is generally set up as a *master station* and the modems at the terminals as *slave stations*. This means that the computer must originate any communications exchange with a terminal; the latter may only respond to signals from the computer. This arrangement, which is natural in a centralized, that is, single-computer-controlled network, gives rise to a line discipline that is different from the idealized contention scheme previously discussed, viz., *polling*.

A typical polling scheme operates as follows. Suppose that there are N terminals. The computer first sends out a character sequence on the line that is equivalent to the question, "Terminal 1, do you have any imput messages to send?" Thus the sequence contains at a minimum an address consisting of a character or two that identifies terminal 1. Whereas all terminals on the line

---

[f]The method of finding ρ varies with the line discipline. For example, in our illustration, the single value of ρ depends on both the input and output message volumes, whereas in the duplex case there are separate values of ρ for each direction of transmission.

Centralized Computer Networks 119

receive this polling inquiry, only terminal 1 responds, usually with another short character sequence. Its response may be that it has no input message to send; that it has one or more such messages to send; or that it is temporarily out of service. In the first instance, the computer will usually send to terminal 1 any output message(s) it has for this terminal[g] (there may be none) and then continue its polling. If terminal 1 has an input message to send, the computer will normally send this terminal a go-ahead character sequence; after receiving the terminal's message(s) and sending an acknowledgment, the computer sends any output messages it has to terminal 1 and continues polling. If terminal 1 reports itself out of service, the computer will continue polling with terminal 2, perhaps after recording the status of terminal 1. There are variations in these methods, for example, if a terminal does not reply to the computer's polling character sequence after a prescribed time interval, the computer may construe this as a terminal out-of-service response. In any event, the computer polls terminals 2, 3, ..., N in this fashion and then starts over with terminal 1; the numbering of the terminals here is not necessarily related to position on the line.

Polling differs from contention in two ways [1]. First, the response time at each terminal is affected by the intervals required for polling. For example, if the line speed is $C = 2,400$ bps and each polling exchange, consisting of the computer's inquiry and a terminal's response, is four characters long, then the time to transmit the polling exchange is 13.3 ms, assuming eight bits per character. In addition, for half-duplex lines there is the *turnaround time* for each modem, that is, the interval that the modem requires to go from transmit to receive or vice versa. This may be as small as 5 to 10 ms for "fast-poll" modems or up to 200 ms for other modems at this line speed. For a hypothetical modem turnaround time of 100 ms and two such times per terminal polled, we have a total polling exchange of 213 ms per terminal. (We neglect here the line propagation time and the computer service time for polling sequences.) As we will see, the polling exchange time for a *successful* poll, that is, one in which the terminal has an input message waiting, increases all service times, whereas the polling exchange time for an *unsuccessful* poll, that is, one in which the terminal has no input message waiting or an out-of-service response, increases the service times for a terminal that must wait for such polls of terminals ahead of it in the polling sequence.

The second way in which polling differs from contention is that the queue discipline of the latter, usually first-in, first-out (FIFO), is replaced by the cyclic polling of terminals in numerical order. If we are interested only in

---

[g]In some message-handling systems, an output message is not necessarily a response to the current imput message. It may be a response to an input message received some time previously, or it may have no logical relationship to any input message.

average delays, this has no effect; for example, as we have noted, the Pollaczek-Khintchine formula, equation 8.3, applies without respect to the queueing discipline. If we wish to estimate other delay statistics such as the variance, we must take this difference into account.

In the previous subsection we analyzed as an example a contention system with $N = 10$ terminals on the line, $C = 2,400$ bps, and so forth. To extend this analysis to the case of polling, let us assume for simplicity that each polling exchange, no matter what the terminal response—no input message, input message waiting, or out-of-service—occupies an average of $t_p = 0.2$ seconds. Then the average service time is increased in two ways. First, it is 0.2 seconds greater than in the contention example by reason of the polling exchange time for the particular terminal whose newly arrived message we are considering. Second, if we assume that when this message arrives, it is the only message waiting for the use of the line, the service time is an additional $(N-1)t_p/2 = (10-1) \times 0.2/2 = 0.9$ seconds greater because, on the average, $(N-1)/2$ terminals that have no message waiting must be polled before the terminal that we are looking at.[h] Thus the average service time with polling is $1.033 + 0.2 + 0.9 = 2.133$ seconds. If we assume that the variance of the polling exchange times is negligible compared to the previous value of $v_s = 0.9$ seconds-squared, we find that the utilization factor for a message-arrival rate of $\lambda = 0.25$ per second is $\rho = 0.25 \times 2.133 = 0.533$ and the average delay, using equation 8.3, is

$$0.533 \times 2.133 \times \frac{[1 + (0.9/2.133^2)]}{2 \times (1 - 0.533)} = 1.458 \text{ seconds}$$

The average response time is $2.133 + 1.458 = 3.591$ seconds.

This model must be modified if the times for the various kinds of polling exchanges are different. In any event, the average time per successful exchange is added to the average service time, as well as $(N-1)/2$ times the average time for an unsuccessful exchange. Martin describes in greater detail the components that enter into both types of polling exchange times, the effects of variations in the line discipline on the average delay, and the estimation of a factor less than $(N-1)/2$ for the average number of unsuccessful polls that precede a particular terminal's poll under certain conditions [1].

---

[h]Martin notes that the assumption that only one newly arrived message is waiting is suitable for the low to moderate line utilizations (ρ less than about 0.6) that generally prevail in practical networks [1]. Also he notes that the queues for the line tend to be dominated by the longer output messages, which reinforces the assumption of only one message waiting.

Practical Aspects

If the message lengths, computer-service times, and message-arrival rate per terminal are fixed, the variables that affect the response time in polling networks are evidently N, the number of terminals on the line; C, the line speed; and the line discipline. Usually there is an upper limit on N imposed by the communications common carrier as a result of line equipment characteristics; for example, in lines up to 9,600 bps in speed N may be limited to 15 or so. (Multipoint lines have rarely been implemented at higher speeds.) Generally duplex operation results in response times that are less than those with half-duplex operation, but the effects of the line discipline must also be considered. Frequently a line is operated with N less than the common-carrier limit, or less than our models might allow, in order to increase overall network reliability since fewer terminals are then affected by a line failure, or to allow a reserve for unexpectedly large message-arrival rates or message lengths.

8-3 Selecting Locations for Multiplexers/Concentrators

In figure 8-1 the locations of the terminals and the computer are generally given. The network design problem is to find the layout of communications lines to connect the terminals to the computer that meets a given performance requirement, for example, one for average response time, and that also minimizes the total communications costs. In chapter 2 we describe two design alternatives for the nonhierarchical network, figure 8-1(a). The first of these, in which there are essentially no constraints on performance, is the minimum spanning tree (MST); it is implemented as one or more multipoint lines, that is, as a tree or as two or more segments. The second alternative is the constrained minimum spanning tree (CMST), that is, a tree or segments in which certain constraints are met such as one on the maximum number of terminals per segment or on the total message-arrival rate per segment. These constraints assure that the performance requirement, for example, for response time, is also met, as described in section 8-2.

As we have noted, the nonhierarchical network, usually the CMST, is not necessarily the minimum-cost solution to the problem. In particular, in many situations we wish to compare the cost of this solution with that of a hierarchical network, as in figure 8-1(b). In the latter we must determine how many concentrators there are (we will use the term *concentrator* from here on rather than "multiplexer/concentrator" for brevity; other general terms such as "access node" are also used); where they are; and how they are connected to the computer, for example, whether by point-to-point lines in a star configuration or with multipoint lines. Further, we must determine for each terminal the concentrator to which it is connected and the method of connection, the most common alternatives again being a star at each concentrator, as in figure 8–1(b), or multipoint lines at each concentrator. It is possible to connect a terminal to two or more concentrators to increase the reliability of the network. We will allow only a single connection per terminal.

Owing to the many possible choices in networks of practical size, we cannot try exhaustively all possibilities, and in fact there are optimum or true minimum-cost solutions only for certain cases. For example, if contrary to our general problem statement, the concentrator locations are given a priori and if the terminals are to be connected in star fashion (point to point) to concentrators, then the optimum terminal-to-concentrator connections can be selected by techniques for solving the *transportation problem* of operations research [3]; the typical constraints are that each terminal is connected to just one concentrator and that no more than a fixed number K of terminals are connected to any concentrator. For convenience of later reference, we will call this the *terminal assignment* problem. In other cases, heuristic algorithms are used, as we will exemplify below.

The Drop Algorithm

Suppose that there are N terminal locations $T_1, \ldots, T_N$ and $M + 1$ possible concentrator locations $S_0, S_1, \ldots, S_M$. Of the latter, S_0 is a particular "concentrator" location, viz., the location of the computer. We wish to select a subset of the concentrator locations such that the total communications costs to connect the terminals to S_0 or to concentrators and the concentrators to S_0 are minimized. All connections of terminals to concentrators or to S_0 and of concentrators to S_0 are point to point or star configurations. Also we require that each terminal be connected either to S_0 or to just one of the concentrators and that no concentrator can have more than K terminals connected to it; the computer may have any number of terminals connected to it. In our illustration of the algorithm, we will show the concentrator locations S_j common with certain of the terminal locations T_i, but this restriction is not necessary.

We may formulate this problem as follows [4]. Let f_j be the cost of concentrator j, including the cost of its communications lines to S_0. Let c_{ij} be the cost of connecting terminal i to concentrator j. We further define two sets of variables that may take the values 0 or 1, that is $x_{ij} = 1$ if T_i is connected to S_j and 0 otherwise, and $y_j = 1$ if there is a concentrator at S_j and 0 otherwise. In the definitions of f_j, c_{ij}, x_{ij}, and y_j, i runs from 1 to N and j from 0 to M; we will assume that $f_0 = 0$ and $y_0 = 1$. We wish to select the x_{ij} and the y_j so as to minimize the cost

$$Z = \sum_{j=0}^{M} y_j f_j + \sum_{i=1}^{N} \sum_{j=0}^{M} x_{ij} c_{ij}$$

subject to the constraints

$$\sum_{j=0}^{M} x_{ij} = 1, \quad i = 1, 2, \ldots, N$$

that is, each terminal connected to just one concentrator or to S_0, and

$$\sum_{i=1}^{N} x_{ij} \leq K, \quad j = 1, \ldots, M \text{ and } y_j = 1$$

that is, no more than K terminals per concentrator in use.

Whereas there are operations-research techniques to solve this *zero-one integer programming problem*, in network applications it is more common to use certain heuristic algorithms. One of these, the drop algorithm, is as follows. We start with all M concentrators in use, and find the minimum-cost configuration by solving the terminal assignment problem for $M + 1$ known concentrator locations $S_0, S_1, \ldots, S_M$. This establishes an initial minimum cost, Z_0^*. In the next step, we drop or delete one concentrator at a time, starting with S_1. In the first part of this step, we imagine that concentrator S_1 is taken out of use, and thus any terminals that had been initially connected to it must now be connected to S_0 or to concentrators $S_2, \ldots, S_M$. We now find a new minimum cost by solving the terminal assignment problem for M known concentrator locations $S_0, S_2, \ldots, S_M$; let us call this minimum cost Z_1^{1*}. We next restore S_1 and the connections of those terminals that had been assigned to S_1 and drop S_2; let us call the new minimum cost Z_1^{2*}. We continue in this way, dropping one concentrator at a time and repeatedly solving the terminal assignment problem. Suppose that we find, after dropping each of the M concentrators in turn, that dropping a particular one, r, results in the greatest cost reduction compared to Z_0^*; let us call the cost Z_1^*. In the next step, we try dropping all combinations of two concentrators at a time, one of them always being r, giving us a new minimum cost Z_2^* and a pair of dropped concentrators r, s. We continue with steps in which we drop all possible sets of $3, \ldots, M$ concentrators, always including the set of k dropped concentrators when we advance to $k + 1$; of course, with M concentrators dropped, all terminals are connected directly to S_0. In practice, the successive minimum costs with $1, 2, \ldots$ concentrators dropped, $Z_1^*, Z_2^*, \ldots$ may not continue to decrease and the algorithm stops. In any event, the solution is the configuration for which the cost is least among $Z_0^*, \ldots, Z_M^*$.

The heuristic aspect of the drop algorithm is the assumption that at any step a true minimum-cost configuration is found by retaining the prior smaller set of dropped concentrators. This is not necessarily true, but the algorithm has nonetheless been successful, as we will discuss further.

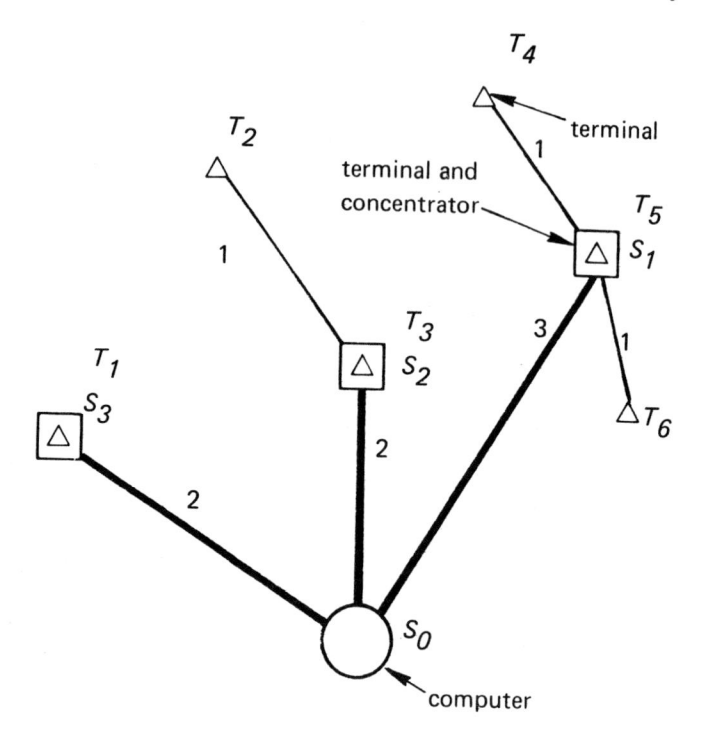

Figure 8-2. Initialized Network for Drop Algorithm

To illustrate the drop algorithm, figure 8-2 shows the initialization of a network in which there are $N = 6$ terminals and $M = 3$ possible concentrator locations.[i] We assume the following costs: $f_1 = 3, f_2 = 2$, and $f_3 = 2$, and

$$c_{ij} = \begin{bmatrix} 1 & 4 & 1 & 0 \\ 2 & 3 & 1 & 2 \\ 1 & 1 & 0 & 2 \\ 4 & 1 & 1 & 1 \\ 4 & 0 & 2 & 2 \\ 2 & 1 & 2 & 3 \end{bmatrix}$$

For example, the cost of connecting terminal T_3 to concentrator S_1 is $c_{31} = 1$. Also, we allow no more than $K = 3$ terminals per concentrator. The

---

[i] In this simple example, it suffices to scan through the c_{ij} matrix row by row and pick out the lowest-cost connection for each terminal, or one of these if there are ties. In larger networks it may be necessary to use a formal solution of the terminal assignment problem.

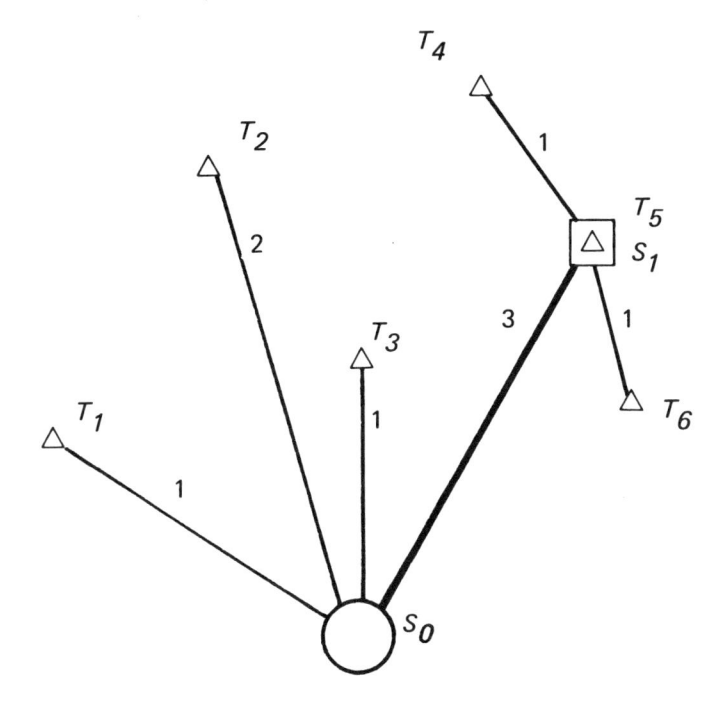

Figure 8-3. A Minimum-cost Solution for the Network of Figure 8-2 with One Concentrator

cost of the initial network is $Z_0^* = 10$; in contrast, the cost of the star connection of all terminals to S_0, with no concentrators in use, is 14. By pursuing the drop algorithm, we find two solutions with two concentrators and $Z_1^* = 9$ and two with one concentrator and $Z_2^* = 9$; one of the latter is shown in figure 8-3.

An Alternative Algorithm for the Star-Star Configuration

We refer to the network illustrated in figure 8-1(b) and discussed in the previous subsection as the *star-star* configuration since both the terminal/concentrator subnetworks and the concentrator/computer subnetwork are made up of point-to-point lines.

An alternative heuristic algorithm for minimizing the communications cost for this configuration is given by Martin [1]. In the first phase of this algorithm, we assume that a concentrator may be at any terminal location—it is not difficult to extend the algorithm to the case in which some or all of the potential concentrator locations are distinct from the terminal locations—and we calculate for each pair of terminal locations T_i and T_j the quantity

$$d_{ij} = \max\left[c_{i0} - \left(c_{ij} + \frac{f'_j}{K}\right), 0\right]$$

Here c_{i0} is the cost of connecting terminal i directly with the computer location S_0 and c_{ij} is the cost of connecting terminal i to a concentrator at location j, as in the previous subsection; f'_j is the part of the concentrator cost at S_j that is for communications between S_j and S_0; and K is the maximum number of terminals that can be connected to a concentrator, as before. Thus d_{ij}, if it is positive, is an estimate of the communications cost savings that result from connecting terminal i to concentrator j; otherwise, $d_{ij} = 0$. Next we find for each terminal location the total savings D_j that can be achieved by connecting terminals to a concentrator at j, that is, the sum of the d_{ij} over those K terminals for which d_{ij} is the greatest, in descending order. We now place the first concentrator at that location j^* for which D_{j^*} is a maximum among all locations $1, \ldots, N$. We continue by finding the second concentrator location in similar fashion, leaving aside the first concentrator and the terminals assigned to it. The algorithm continues in this way until all terminals have been assigned to concentrators.

When the foregoing first phase has ended, we have a reasonable set of concentrator locations, approximately N/K in number. To improve the results, the algorithm now goes into a second phase in which each terminal i is examined in turn {Martin suggests that this start with the terminal i that is most distant from S_0, or in general, for which c_{i0} is greatest [1]} and its assignment is changed if either of the following is true: (a) it is assigned to concentrator r and there is another concentrator s with spare capacity, that is, with fewer than K terminals connected to it, with $d_{is} > d_{ir}$; or (b) it is assigned to concentrator r and it would cost less to connect it directly to the computer center, that is $c_{i0} < c_{ir} + f'_r/K$. Of course, we choose the lower cost reassignment between (a) and (b).

This algorithm is somewhat similar to another one called the add algorithm [4] in that concentrators are added one at a time. Both algorithms are thus inverses, in a sense, of the drop algorithm. In a large network, the end results for the three algorithms are not necessarily the same as to concentrator placement and thus as to cost, although this is the case for simple networks such as that in figure 8-2.

Since f'_j is only the communications cost for a concentrator at j, we must add to the communications costs from Martin's algorithm the concentrator equipment costs to find the total system costs.

Algorithms for Other Configurations

The principal configurations other than the star-star that have been subjected to cost minimization are (a) those in which the terminals are in stars connected to the concentrators and the concentrators are connected on multipoint lines, that is, on a tree or segments, to the computer; and (b) conversely, those in which the terminals are connected on multipoint lines to the concentrators and the concentrators are in a star centered at the computer. Martin describes an algorithm for case a in which the concentrator locations are found first, using a slight modification of the algorithm described in the previous subsection, and then the multipoint line to connect the selected concentrator locations to the computer is designed with the Esau-Williams algorithm (see chapter 2). In the latter step, the constraints on the multipoint line arise from the computer's ability to handle the total number of concentrators and total message rate, and from the line utilization. In the balance of this subsection we will deal with case b, which has attracted more interest in practice.

Several heuristic algorithms for case b have been published and used, since no theoretically optimum method is known. Martin describes an algorithm in which the first step is to estimate the communications cost savings that accrue from putting a concentrator at each location i in turn [1]. (These savings are estimated, not calculated, because the layouts of the multipoint lines for the terminals are not yet known.) In this step, terminals are assigned to potential multipoint lines at the concentrator at i by a search procedure that scans outward from i, adding terminals subject to remaining within the message-load constraints. Concentrators are placed at the locations that have the greatest potential savings. Finally, the multipoint lines for the terminals are designed by the Esau-Williams algorithm.

Other algorithms for case b rely on different means of *clustering* the terminals, that is, of partitioning the terminals into subsets or clusters such that the terminals in each subset can ultimately be profitably connected on the same multipoint line to a concentrator. The latter is usually at one of the locations in the subset, but it may be at another point that is close geographically or in cost to the terminals in the cluster. Representative papers that contain complete algorithm descriptions are those of McGregor and Shen [5] and of Dysart and Georganas [6]. In the former, clusters are formed of nearby terminals, with the terminals being added so long as the message load for the totality of terminals does not exceed the constraint. In the latter, clusters are formed by assigning each terminal to the nearest concentrator, the concentrator locations having been chosen first by a technique that favors a high number of appearances on the nearest-neighbor lists of other locations. It is evident from these papers and others that the clustering method is the critical part of the algorithm. Once the clusters that tend to generate the minimum communications cost are identified, the multipoint line layouts at each concentrator are carried out by well-known algorithms such as Esau-Williams.

References

1. Martin, J. *Systems Analysis for Data Transmission*. Englewood Cliffs, N.J.: Prentice-Hall, 1972.
2. Syski, R. *Introduction to Congestion Theory in Telephone Systems*. Edinburgh: Oliver and Boyd, 1960.
3. Boorstyn, R.R., and Frank, H. "Large-scale Network Topological Optimization," *IEEE Trans. Communications* COM-25(1):29–47, 1977.
4. Schwartz, M. *Computer-Communication Network Design and Analysis*. Englewood Cliffs, N.J.: Prentice-Hall, 1977.
5. McGregor, P.V., and Shen, D. "Network Design: An Algorithm for the Access Facility Location Problem. *IEEE Trans. Communications* COM-25(1):61–73, 1977.
6. Dysart, H.G., and Georganas, N.D. "NEWCLUST: An Algorithm for the Topological Design of Two-level, Multidrop Teleprocessing Networks," *IEEE Trans. Communications* COM-26(1):55–62, 1978.

9 Least-cost-route Selection

9-1 Introduction

In common-user telecommunications systems with circuit switching such as private branch exchanges (PBXs), there are generally several kinds of trunks over which users make outgoing calls. These include *local exchange trunks*, which may also be used for *Direct Distance Dialing* (DDD, or toll) calls; *Wide Area Telecommunications Service* (WATS) trunks,[a] for calls to designated remote areas; *foreign exchange* (FX) trunks, for calls to a specific remote exchange; and *tie lines* for calls to a remote PBX. Ideally, we wish, as each call is placed by a PBX user, to select for it the type of trunk that will minimize the *total monthly trunk costs*,[b] at the same time maintaining an acceptable quality of service. We may imagine a "black box" interposed between the PBX and the outgoing trunks, as in figure 9–1, that receives the dialed telephone number for each call; uses an algorithm that depends on the call destination to choose the optimum kind of trunk; and connects the call to such a trunk.

We cannot achieve this ideal in practice, since the optimum choice of trunk type for each call depends on the numbers and destinations of calls yet to be made later in the month. Instead, we estimate the average monthly call volumes and holding times for each destination, and we design the algorithm for the black box to minimize the total monthly trunk costs if these data were to prevail. Although there are inevitable random and perhaps seasonal fluctuations in the traffic quantities, it is possible in this way to obtain useful solutions to the problem, that is, to reduce monthly costs at a given total call volume and quality of service compared to those in a nonoptimized network. This is the problem of *least-cost-route or -trunk* selection, also called by other names such as *automatic route selection* and *telephone call management*. In this chapter we state the general problem of least-cost routing and give solutions for a limited but practically sufficient version of this problem for several call-handling methods.

---

[a] As of 1980 there were impending changes in interstate WATS as to the nature of the service and the rates. In this chapter WATS represents any message toll service characterized by lower hourly charges at high levels of monthly usage.
[b] The monthly interval is the common one for telephone billing in the United States.

PBX users

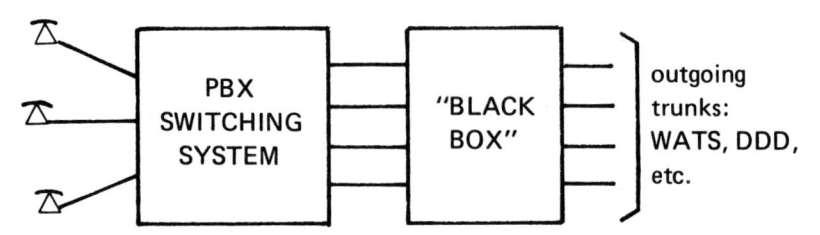

Figure 9-1. A "Black Box" for Least-cost-route Selection

Note that the "black box" might alternatively be integral to the PBX switching system, or it might be a remote computer.

9-2 Problem Statement

The selection of the trunk type for a given call is highly dependent on the common carrier's *rate structure* for each type. Indeed, it is these differences in rates that give rise to the problem in the first place. In the United States, for example, DDD trunks bear a relatively small fixed monthly charge, and there is an additional charge for each customer-dialed toll call that is based on its duration, the time of the day, day of the week, and distance. A WATS trunk, on the other hand, has a larger fixed monthly charge than a DDD trunk for monthly usage up to a fixed number of hours; usage over this amount per month is charged at another rate. Both of these rate components are independent of time, day, and distance provided that the called destination is in the designated WATS-coverage area. FX trunks and tie lines have yet other rate characteristics.

Suppose that we have M different types of trunks—DDD, WATS, and so on. A simple method of trunk selection is to partition the calls by destination into M groups and to complete calls in each group only over trunks of one type. For example, a call to a destination in WATS band 1[c] is completed only over a band 1 WATS trunk, and if all trunks in this group are busy, the call is blocked. This is not necessarily an economical arrangement, since it may cost less, for example, for a given probability of blocking to have calls that are blocked in the band 1 WATS group *overflow* to a group of band 2 WATS trunks, or to a group of DDD trunks. As another example, it is

---

[c]In the interstate tariffs of the Bell System companies in the United States, this means an area contiguous to the user's own state but not including it. Higher numbered bands consist of successively larger areas, each one including all lower numbered bands. The monthly rates increase with the band number.

Least-cost-route Selection

desirable to complete all calls to a location served by tie lines over the appropriate tie-line group, but it may be less costly to have a small proportion of blocked tie-line calls overflow to DDD, and so on, than to install extra tie lines. We would therefore like to prescribe not only the numbers of trunks of each of the M types but also the overflow sequences. These might be quite complex since, for example, calls whose first choice is a trunk of type A might overflow first to type B, and then from there to type C.

Fortunately, in many practical systems it suffices to deal with a more limited problem, as follows. We are given the daily and hourly numbers and durations of calls for a month by destination. We assume that these calls will be completed on a group of N trunks, of which K_i are of type i, $i = 1, 2, \ldots, M$; $K_1 + K_2 + \cdots K_M = N$. Furthermore, the N trunks are operated as a single full-access group[d] in such a way that the K_1 trunks of type 1 are tried first, and if these are all busy, the call seeks a connection among the K_2 trunks of type 2, and so on. Thus each successive subgroup i overflows to subgroup $i + 1$, $i = 1, \ldots, M - 1$. Calls that find all N trunks busy are not served. We wish to select N and the K_i so that the total monthly trunk costs are minimized.

That the solution of this limited problem is the key to the general problem can be seen as follows. If we fix the level of traffic offered to a full-access trunk group during some period such as the busy hour and observe the usage on successive trunks 1, 2, ..., we find that trunk 1 is used the most; trunk 2 less than trunk 1; and so on up to the highest numbered trunk in the group, which is used the least.[e] By extension, this property carries over to longer periods such as the month. Furthermore, the monthly costs per trunk for most common trunk types are of the form $F + VU$, where F is the fixed monthly charge per trunk, V is the additional charge per hour of usage, and U is the usage in hours per month. As we have noted in the example of WATS, cost F may include a fixed number of monthly hours of usage, and thus U is the usage in excess of this amount per month. In figure 9–2 we show the monthly costs per trunk versus usage per month for two types: type A has a smaller value of F than type B, and its incremental cost VU is incurred starting at $U = 0$, whereas type B has an incremental cost only for usage more than the threshold value U_T. We note that a trunk of type A costs less per month than one of type B up to a monthly usage of U_0 hours, whereas for more than U_0 hours per month a trunk of type B costs less. Thus if our choice for a trunk is limited to type A or B, we would choose the former if the usage is U_0 or less and the latter if it is more than U_0. If we now put this observation together with the description of tapered usage of trunks in a full-access group,

---

[d] See chapter 5. We use the term *full access* interchangeably with *full availability*.
[e] We will look at these relationships quantitatively later in this chapter.

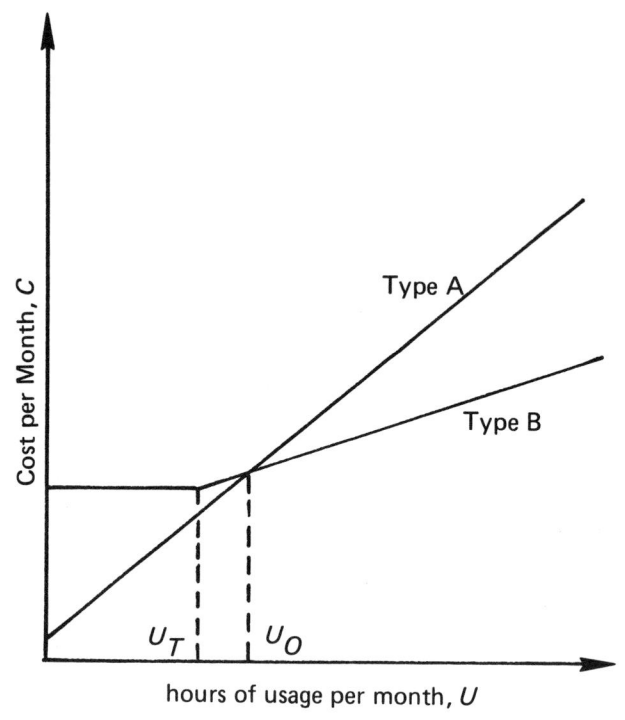

Figure 9-2. Monthly Cost versus Usage for One Trunk, Two Different Types

as given earlier, we find that in a full-access group composed only of types A and B, a subgroup of trunks of type B is the first choice; once we reach a trunk number high enough so that the usage falls below U_0 hours per month, the balance of the trunks are of type A. This principle extends to three or more types of trunks. A common progression of three types in practice, for example, is that the first-selected trunk type is Full Business Day WATS, which has a higher F value, higher threshold value U_T, and lower V value than Measured WATS; the second trunk type is Measured WATS; and the third type is DDD.[f]

In practice, we may have both a full-access group, consisting of diverse trunk types as just described, and one or more special trunk groups with a separate access pattern. An example of such a special group is a tie-line group that is always selected first for calls of this one type. If all of the tie

---

[f]If the monthly usage on the lowest numbered trunks is not enough to justify Full Business Day WATS, then the progression is from Measured WATS to DDD. If the monthly usage on the lowest numbered trunk falls below the crossover U_0 between DDD and Measured WATS, it is economical simply to use all DDD trunks, and there is no least-cost-route selection.

lines are busy, the call may be allowed to overflow to the main full-access group and to progress through WATS, DDD, and so on, subgroups. This might be allowed if the value of call completion for high-priority users justifies the possible extra cost per call.

In the following subsections we examine in more detail the properties of a full-access trunk group under various options for the disposition of calls that are not completed at once.

9-3 The Full-access Trunk Group with Blocking

We have dealt in chapter 5 with trunk groups with *blocking* or *clearing*, in which a call that does not immediately find an idle trunk is denied service. The caller receives a *busy signal*, and the call is assumed to disappear from the system.

Under the assumptions that support the Erlang B formula, explained in chapter 5, the traffic carried on trunk number m of a full-access group of c trunks ($m = 1, 2, \ldots, c$) is [1]

$$a_m = a[B(m-1, a) - B(m, a)] \tag{9.1}$$

Here a is the traffic offered to the group of c trunks, in erlangs, and $B(x, a)$ is the Erlang B formula for x trunks and a erlangs of offered traffic; we recall the convention that $B(0, a) = 1$. For example, suppose that we have $c = 5$ trunks and $a = 3.0$ erlangs of offered traffic. From equation 9.1 we find that the traffic carried on each trunk is as follows:

m	a_m
1	0.75
2	0.66
3	0.55
4	0.42
5	0.29
Total	2.67 erlangs

We note that this exemplifies the pattern of decreasing usage versus m mentioned earlier. Also the blocked traffic is $3.0 - 2.67 = 0.33$ erlangs, which corresponds to a blocking probability of $B(5, 3.0) = 0.11$. Tables are available that give a_m versus m over a range of useful values of a and c [2, 3].

Equation 9.1 holds, as does the Erlang B formula upon which it is based, during a period such as the busy hour in which the offered traffic a remains constant. To apply equation 9.1 to the estimation of *monthly* usage per trunk, it is customary to assume that offered traffic is constant at a_{ij} erlangs during hour i ($i = 1, 2, \ldots, 24$) of day j ($j = 1, 2, \ldots, 30$, and so on), or to otherwise subdivide the month into periods of approximately uniform offered traffic. Then equation 9.1 is applied separately for each period, that is, for each value of offered traffic a_{ij} and the resulting amounts of usage on each trunk $m = 1, \ldots, c$ are added up. For example, if the whole month's traffic consisted of just one hour per day in which the group of five trunks were offered 3.0 erlangs, and if there were 22 business days per month, the usage on trunk 2 would be $22 \times 0.66 = 14.5$ hours per month. In general, the monthly usage on trunk m in hours is

$$\alpha_m = \sum_i \sum_j a_{ij}[B(m-1, a_{ij}) - B(m, a_{ij})]T_{ij} \qquad (9.2)$$

in which T_{ij} is the time interval, in hours per month, during which the offered traffic is assumed to be constant at a_{ij} erlangs.

In engineering a full-access trunk group, we first size the group according to the desired blocking probability in the busiest hour. For example, if the busy-hour offered traffic is 3.0 erlangs, a group of $c = 5$ trunks is large enough if we can tolerate a blocking probability of 0.11; otherwise we would consult a table such as table 5–1 and choose a higher value of c, corresponding to a lower blocking probability.

To subdivide the trunk group for least-cost routing, we use data such as those contained in figure 9–2, that is, the cost per month versus hours of usage per month for one trunk of each type. We start with trunk 1 and assign to it the type for which the monthly cost is least for α_1 hours of usage per month. We continue in this way, selecting the least-monthly-cost type for trunks $2, \ldots, c$. As we have noted, this will usually result in a first-choice subgroup of which all trunks are of the same type, followed by a second-choice subgroup all trunks of which are of another type, and so on.[g]

In estimating the busy-hour traffic a_{ij} we must include in the trunk holding time for each call not only the conversation time but also the time it takes for the common-carrier switching system to set up the connection and for the called party to anwer since the trunk is occupied during these intervals to the exclusion of other calls.

---

[g]This assignment of trunk types is optimum for the assumed hourly and daily offered traffic values a_{ij}. We may therefore need to adjust the trunk type assignments from time to time as the monthly traffic data change, for example, in response to seasonal business changes. Note also that to be strictly correct, we may need to take into account the time and day of each call in estimating trunk costs, as in the case of a DDD trunk. Since most business calls are made during the day hours of weekdays, this requires only a small correction for out-of-hours calls.

Least-cost-route Selection

Repeated Call Attempts

In section 5-3 we describe a method of calculating blocking probabilities in which we account for the tendency of telephone-system users to reattempt their blocked calls. In this method we assume that the total offered traffic, a_R, in the busy hour consists of the original or first-attempt offered traffic a erlangs increased by a quantity Δa that is a proportion ρ of the total blocked traffic, that is, the blocked traffic when the offered traffic is $a_R = a + \Delta a = a + \rho a_R B(c, a_R)$, as in equation 5.7. We point out in section 5-3 that this equation can be solved recursively for a_R, given ρ; the initial offered traffic, a erlangs; and the size of the trunk group, c. If $\rho = 0$, there are no reattempts of blocked calls, and the blocking probability is $B(c, a)$; if $\rho > 0$, some calls are reattempted, and more of the initial offered traffic a is carried at the penalty of a higher blocking probability $B(c, a_R)$.

As noted in section 5-3, this method relies on the assumption that the totality of call attempts, both original and repeated, has a Poisson distribution of arrivals. This amounts to assuming that a user's reattempts of a particular call are not correlated in time with the original call, or in other words that they occur after an idle interval of a few average holding times, rather than rapid succession just after the original attempt. This has been described as the assumption of *slow retrials* [4].

Provided that this assumption holds, we may apply equation 9.1, with a_R substituted for a, to find the traffic carried in each trunk of the group. For example, if $a = 3.0$ erlangs and $c = 5$ trunks, as in our previous example, and if $\rho = 1$, that is, all initially blocked calls are reattempted and eventually carried, we find from the recursive solution of equation 5.7 that $a_R = 3.674$ erlangs, and the a_m from equation 9.1 are as follows:

m	a_m
1	0.782
2	0.710
3	0.618
4	0.507
5	0.383
Total	3.000 erlangs

We notice that each trunk carries somewhat more traffic than for the no-reattempt case, $\rho = 0$. These increases in turn mean that we may assign more economical types of trunks for least-cost-route selection and achieve a lower total monthly trunk cost. This is reasonable since the blocking probability

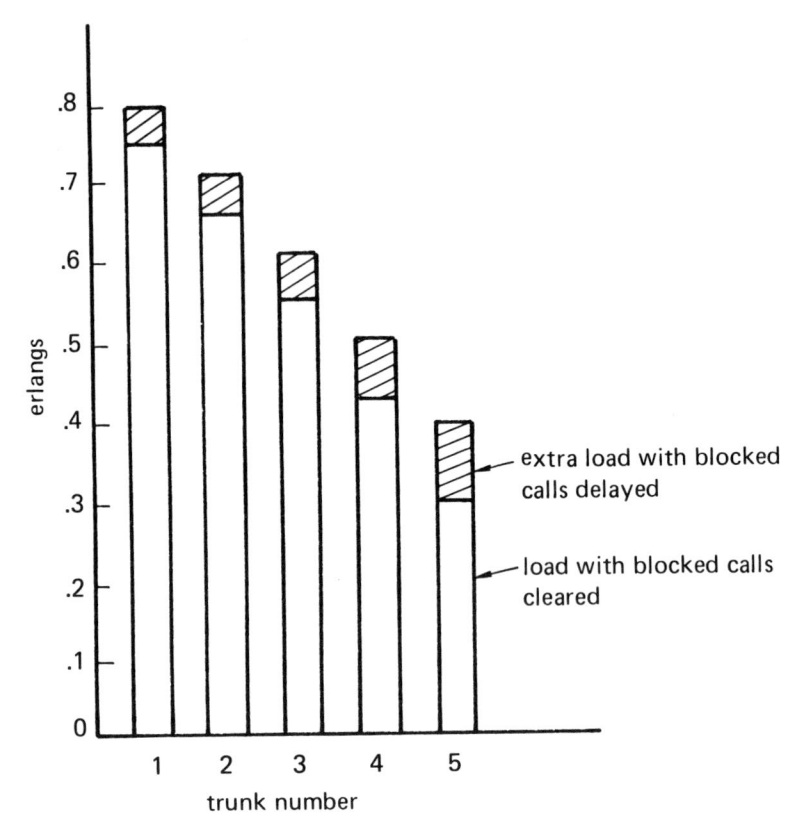

Figure 9-3. Carried Traffic or Load per Trunk with $c = 5$ Trunks and $a = 3.0$ Erlangs

with 100 percent reattempts is $B(5, 3.674) = 0.16$, as compared to the original value of $B(5, 3.0) = 0.11$.

Recent literature refers also to models for *fast retrials*, in which reattempts occur after a smaller interval, one average holding time or less [4].

9-4 The Full-access Trunk Group with Delays

An alternative way to operate a full-access trunk group is with *delays*. Here, if there is no idle trunk in the group when a call is attempted, the call waits in a queue until a trunk becomes available and is then served.

With the assumptions of Poisson arrivals of calls and a negative exponential distribution of holding times, the probability that a call will encounter a delay is given by

Least-cost-route Selection

$$C(c, a) = Y/\left(Y + \sum_{i=0}^{c-1} a^i/i!\right) \tag{9.3}$$

where, as before, a is the offered traffic in erlangs, and c is the number of trunks, and also $Y = a^c c/c!(c-a)$. Equation 9.3 is called the *Erlang C* formula [1], which has also been tabulated [5]. It is the multiserver ($c > 1$) extension of the single-server delay model that we have used in chapters 6 and 7. It is customary to assume a first-in, first-out (FIFO) discipline for the queue, but this is not germane to our application here. Also we assume that the system has enough capacity to retain waiting calls that only rarely is this capacity exceeded.

In equation 9.3 we must have $c > a$ to avoid the buildup of a queue of arbitrarily large size. In a delay system that meets this practical restriction, all the offered traffic is eventually carried, with increasing delays as a approaches c from below. If the average holding time is h seconds, the probability of a delay greater than t seconds is

$$P(>t) = C(c, a) \exp\left[-(c-a)t/h\right] \tag{9.4}$$

and the average delay on all calls is

$$d = \frac{hC(c, a)}{(c-a)} \tag{9.5}$$

For convenience of calculation, we note that

$$C(c, a) = \frac{B}{[1 - u(1 - B)]} \tag{9.6}$$

where B stands for $B(c, a)$ and $u = a/c$. Thus if a table of the Erlang B function is at hand, the values of the Erlang C function may be calculated easily.

The analog of equation 9.1 for finding the usage, b_m, on trunk m ($m = 1, \ldots, c$) in a group with delays is [1]

$$b_m = a_m + \frac{a(1 - a_m)C(c, a)}{c} \tag{9.7}$$

where a_m is the usage on the same trunk calculated by equation 9.1. It is evident from equation 9.7 that for given values of a and c, $b_m > a_m$. Thus with delays we achieve higher loading on each trunk than in the case of blocking (without repeated attempts). To continue with our example with

$a = 3.0$ erlangs and $c = 5$ trunks, we find the following values of carried traffic or usage per trunk from equation 9.7:

m	b_m
1	0.785
2	0.709
3	0.614
4	0.503
5	0.389
Total	3.000 erlangs

A proportion $C(5, 3.0) = 0.24$ of all calls experiences some delay, and the average delay on all calls, assuming an average holding time of 3 minutes, is, from equation 9.5, about 21 seconds. Much as in the case of blocking with repeated attempts, we may impose a penalty on the users, viz., delays, in exchange for the lower total monthly costs that are achievable by reason of the greater loading per trunk.[h] Figure 9-3 shows a comparison of the a_m and b_m values.

There are two common modes of operation of trunk groups with delays in automatic and semiautomatic switching systems. With *hold-on* queueing, the user waits with his telephone off-hook until a trunk is available. In contrast, with *call-back* queueing, he dials the called number and, if a trunk is not immediately available, dials a call-back address or command and then hangs up; the switching system calls him back later when a trunk is available. Not surprisingly, people are willing to wait for longer times in the call-back mode than in the hold-on mode since in the former case they may resume their activities rather than "hang on" to the telephone. Thus a higher average delay on all calls d is permissible in the call-back mode, and in turn a higher ratio of a/c, higher loading per trunk, and lower monthly trunk costs under least-cost-route selection.

References

1. Syski, R. *Introduction to Congestion Theory in Telephone Systems*. Edinburgh: Oliver and Boyd, 1960.
2. Mina, R.R. *Introduction to Teletraffic Engineering*. Chicago: Telephony Publishing, 1974.

---

[h]The penalty of delays is imposed by the mode of operation of the switching system whereas the penalty of increased blocking, necessitating repeated attempts, is imposed by furnishing fewer trunks than would be required for a low blocking probability.

3. Goeller, L.F., Jr. "Erlang B Tables and How to Use Them," *Telecommunications* (Handbook and Buyers' Guide issue), 1975.
4. Jewett, J.E. et al. "Systems Approach to Network Design: Choosing the Right Technique," *Bus. Commns. Rev.*, May-June 1980.
5. Descloux, A. *Delay Tables for Finite and Infinite Source Systems*. New York: McGraw-Hill, 1962.

Index

Add algorithm, 122
Aggarwal, K.K., 49
Algorithm(s): add, 122; alternative, for star-star configuration, 125–126; Dijkstra, 27–29; drop, 122–125; Esau-Williams, 14–17, 18, 127, 128; Floyd, 22–27, 29; labeling and augmentation, 38–39, 41, 43; network blocking probability, 80–82; for other configurations, 127–128; Prim, 10–12, 14, 16; route-tracing, 30–31
Alternate routing, 64–74, 97–98
Arc, 3
ARPANET, 103, 106–108
Automatic route selection. *See* Least-cost-route selection

Backbone groups, 65
Backbone network, 85
Bajaj, D., 82–83
Baran, P., 103
Bellmore, M., 49
Blocking: full-access trunk group with, 133–136; probabilities with alternate routing, 67–74; probabilities in networks, 74–83
Branch, 3
Branch-and-bound method, 18
Busy period, 55

Call: blocked, 53, 54; cleared, 54; lost, 54
Call-back queueing, 138
Capacity assignment (CA) problem, 92–98, 108–109
Capacity and flow assignment (CFA), 100
Capacity table, Erlang B, 56
CCITT, 67, 69, 74
Centralized computer networks, 113; estimation of response time in, 113–121; selecting locations for multiplexers/concentrators in, 121–128

Cheung, C.K., 100
Chou, W., 14, 16, 17
Circuit-switched networks, 53–54; and alternate routing, 64–74; and blocking probabilities in networks, 74–83; and full-availability trunk group, 54–60; and trunk group under other assumptions, 60–64
Clustering, 127–128
Cohesion: and connectivity, 35–36; of undirected graph, finding, 36–41
Components, routes, and cut-sets, 4–6
Computer networks. *See* Centralized computer networks
Concentrators, 113; selecting locations for, 121–128
Configuration(s): algorithms for other, 127–128; alternative algorithm for star-star, 125–126
Connection matrix, 6
Connectivity: cohesion and, 35–36; m-, 43–44; network, 97; of undirected graph, finding, 41–44
Constrained minimum spanning tree (CMST), 121
Constraints, trees with, 12–14
Contention model, 115–118
Covo, A.A., 82
Cut, 37
Cut-sets, 37; link, 5; methods based on link, 48–49; node, 5–6; routes, components, and, 4–6
Cycles, trees and, 6

Decomposition, 26
Degree, of node, 3
Delay(s): full-access trunk group with, 136–138; round-trip, 107–108; time, 104, 105
Dijkstra, E.W., 27, 110; algorith, 27–29
Direct Distance Dialing (DDD), 129, 130–133
Disjoint routes, node- and link-, 31–33
Dreyfus, S.E., 29

141

Drop algorithm, 122–125
Dysart, H.G., 127

Edge, 3
Elias, D., 17
Ends, of links, 3
Engset formula, 62
Erlang, A.K., 55
Erlang B formula, 55–60, 62, 67, 133, 137
Erlang C formula, 137
Esau, L.R., 14
Esau-Williams algorithm, 14–17, 18, 127, 128
External flows method, 108

Ferguson, M.J., 17
First-in, first-out (FIFO) queue discipline, 88, 118, 137
Flow assignment (FA) problem, 98–100, 109–111
Flow control, 111
Flow deviation method, 100, 109
Flow pattern, feasible, 37–38
Floyd, R.W., 22; algorithm, 22–27, 29
Ford, L.R., Jr., 38, 110
Foreign exchange (FX) trunks, 129, 130
FORTRAN program, 23, 29
Frank, H., 43
Fulkerson, D.R., 38, 110

Georganas, N.D., 127
Gomory, R.E., 41
Grade of service, 56
Gradient projection method, 100
Grading, 60
Graphs: complete, 36; completely corrected, 9; connected, 5; directed, 36–38, 90; disconnected, 5; finding cohesion of undirected, 36–41; finding connectivity of undirected, 41–44; and networks, 3–4; weights in, 7–8
GTE Telenet and Tymnet, 107

Hänsler, E., 49
Hierarchical network, 113

Holding time, 55
Hold-on queueing, 138
Holtzman, J.M., 82
Hu, T.C., 26–27, 41

Independence assumption, 90
Interarrival times, 54, 87

Jagerman, D.L., 60
Jensen, P.A., 49
Jewett, J.E., 64

Karnaugh, M., 17, 18
Katz, S., 82
Kendall, D.G., 89
Kershenbaum, A., 14, 16, 17
Kleinrock, L., 85, 86; and capacity assignment (CA) problem, 92–98, 108; extensions of model of, to packet-switched networks, 103, 104–108; flow assignment (FA) problem, 98–100; his model of message-switched network, 89–92
Kleitman, D.J., 43–44
Klemushin, G.N., 33
Kruskal, J.B., 16
Kuczura, A., 82–83

Labeling algorithms, 27
Labeling and augmentation algorithm, 38–39, 41, 43
Least-cost-route selection, 129; and full-access trunk group with blocking, 133–136; and full-access trunk group with delays, 136–138; problem statement, 130–133
Lee, C.Y., 47
Line discipline, 115. *See also* Contention model; Polling
Link(s), 3; -blocking probability, 74; -completion probability, 75; -disjoint and node-disjoint routes, 31–33; flow, 37; and node failures, probabilistic, 44–49
Local central office, 53
Local exchange trunks, 129

Index

McGregor, P.V., 127
Martin, J., 15, 118, 120, 125, 126, 127
Max-flow min-cut theorem, 36-38
m-connectivity, 43-44
Meister, B., 96
Message rate, 13
Message-switched networks, 85-86; and capacity assignment (CA) problem, 92-98; and flow assignment (FA) problem, 98-100; Kleinrock's model of, 89-92; and single-server queue, 86-89
Minimum spanning trees (MSTs), 9-12, 13, 18-19, 29, 121; constrained (CMSTs), 121
M/M/1 system, 89
Multicommodity flow problem, 99-100
Multiple-packet messages, 105-106
Multiplexers, 113; selecting locations for, 121-128
Multipoint or multidrop line, 113

Naylor, W.E., 106, 107
Network(s): blocking probabilities in, 74-83; and cohesion, 35-41; and connectivity, 35-36, 41-44, 97; graphs and, 3-4; hierarchical, 113; nonhierarchical, 114; and probabilistic link and node failures, 44-49; reliability, deterministic and probabilistic measures of, 35. *See also* Centralized computer networks; Circuit-switched networks; Message-switched networks; Packet-switched networks
Node(s), 3; degree of, 3; -disjoint and link-disjoint routes, 31-33; failures, probabilistic link and, 44-49; neighbors of, 27; -pair blocking probability, 74; -pair completion probability, 76; pairs, shortest routes for, 22-27; pendant, 26; shortest routes from specified node to all other, 27-29
Nonrandom calls, 62-64

Objective function, 100, 109

Overflow pattern, 130-131

Packet-switched networks, 103; extensions of Kleinrock's model to, 103, 104-108; and network optimization, 108-111
Partial availability, 60
Peakedness, of overflow traffic, 65-67
Peakedness ratio, 66-67; weighted mean, 69
Pendant nodes, 26
Point-to-point communications line, 113
Poissonian input, 54-55, 56, 62, 63
Pollaczek-Khintchine formula, 116-117, 120
Polling, 115, 118-120
Prim, R.C., 9; algorithm, 10-12, 14, 16
Private branch exchanges (PBXs), 129
Probabilistic link and node failures, 44-49

Queueing: cell-back, 139; hold-on, 138. *See also* Single-server queue

Rate structures, 130
Reliability: network, 35; terminal, 44
Repeated call attempts, 62-63, 135-136
Response time, estimation of, in centralized computer networks, 113-121
Retrials: fast, 136; slow, 135
Route(s), 21; acceptable, 30-31; components, cut-sets, and, 4-6; directed, 37; generating longer, 29-31; node-disjoint and link-disjoint, 31-33; shortest, for all node pairs, 22-27; shortest, from specified node to all other nodes, 27-29; -tracing algorithm, 30-31; and weights, 21-22
Routing: adaptive, 98: alternate, 64-74, 97-98; fixed, 90; optimization (*see* Flow assignment problem); random, 98; table, 78

Schwartz, M., 100, 110-111
Segal, 78, 82

Segments, 12
Service time, 87, 104, 105, 116; average, 88
Shek, C.H., 22
Shen, D., 127
Single-packet model, 104–105
Single-server queue, 86–87; assumptions, 87–88; average time in system, 88–89
Speech, packetized, 103
Star-star configuration, alternative algorithm for, 125–126
Station(s): master, 118; slave, 118
Statistical equilibrium, 55, 56
Stern, T.E., 110–111
Store-and-forward switching, 85
Subgraph, 3
Subscriber loop, 53
Suurballe, J.W., 33

Telephone call management. *See* Least-cost-route selection
Terminal assignment problem, 122
Terminal reliability, 44
Threshold network throughput, 108
Throughput, 87
Tie lines, 129
Time in the system, 88, 91
Topology, capacity, and flow assignment (TCFA), 100
Traffic: blocked, 64, 66; carried, 55, 56; high-throughput, 103; low-delay, 103; offered, 55; overflow; 64–68; peakedness of overflow, 65–67; real-time, 103

Traffic engineering, 53–54
Transportation problem, 122
Trees: constrained minimum spanning (CMSTs), 121; with constraints, 12–14; cycles and, 6; minimum spanning (MSTs), 9–12, 13, 18–19, 29, 121; minimum-weight constrained, 14–17, 18, 29; other methods for finding optimal constrained, 17–18; segments of, 12; shortest-route, 29; weight of, 8
Trunk groups, 53; full-access, with blocking, 133–136; full-access, with delays, 136–138; full availability, 54–60; and nonrandom call requests, 62–64; partial availability, 60

Union, of graphs, 3
Utilization factor, 88 and n, 117

Vertex, 3
Vogel, 16

Weights, 21–22; in graphs, 7–8; link, 7–8; node, 8
Wide Area Telecommunications Service (WATS), 129, 130, 132–133
Wilkinson, R.I., 64
Wilkov, R.S., 35
Williams, K.C., 14. *See also* Esau-Williams algorithm

Zero-one integer programming problem, 123

About the Author

Howard Cravis is a senior member of the professional staff at Arthur D. Little, Inc., where he has specialized in communications-systems engineering studies of voice, data, video, facsimile, and composite communications networks. He was previously employed at Bell Telephone Laboratories, where he carried out investigations of speech interpolation, pulse-code-modulation carrier systems, and applications of probability theory to communications problems. He also worked at MITRE Corporation and at GTE Sylvania, where he performed analytic modeling of military communications networks. He received his undergraduate education in electrical engineering at Columbia University, and received graduate degrees from Columbia and from Harvard University.